Estimation of Residual Static Time Shifts by means of the CRS-based Residual Static Correction Approach

————————

Bestimmung reststatischer Verschiebungszeiten mit Hilfe der CRS-basierten Reststatikkorrekturmethode

Zur Erlangung des akademischen Grades eines

DOKTORS DER NATURWISSENSCHAFTEN

von der Fakultät für Physik der

Universität Karlsruhe (TH)

genehmigte

DISSERTATION

von

Dipl.-Geophys. Ingo Koglin

aus

Karlsruhe

Tag der mündlichen Prüfung: 03. Juni 2005

Referent: Prof. Dr. Peter Hubral

Korreferent: Prof. Dr. Friedemann Wenzel

Bibliografische Information Der Deutschen Bibliothek

Die Deutsche Bibliothek verzeichnet diese Publikation in der Deutschen
Nationalbibliografie; detaillierte bibliografische Daten sind im Internet über
http://dnb.ddb.de abrufbar.

ISBN 3-8325-0977-1

Logos Verlag Berlin
Comeniushof, Gubener Str. 47,
10243 Berlin
Tel.: +49 030 42 85 10 90
Fax: +49 030 42 85 10 92
INTERNET: http://www.logos-verlag.de

Abstract

Reflection seismic measurements are used to indirectly investigate the earth's interior to obtain an image of the seismic velocity distribution of the subsurface at the investigated area. This velocity distribution is of great interest for oil and energy companies. With the velocity distribution, it is possible to obtain an image of the geological structures in the subsurface which, in early days, was mainly used to find hydrocarbon reservoirs. Nowadays, it is also important to develop natural energy sources like geothermal energy.

In case of onshore measurements, the recorded reflection events are usually degraded in quality due to traveltime distortions caused by the uppermost layer, i.e., the so-called weathering layer. This name implies that mainly weathering influences its effect on reflection events. Within the uppermost layer the pores are usually air filled which results in a low velocity for this layer. The seismic waves have to traverse this low-velocity layer at least once to emerge at the receivers where they are recorded. The influence of the weathering layer on the reflection traveltimes is usually compensated by static corrections based on the assumption of surface consistency and time invariance. These static corrections can be divided into two parts: the field or datum static correction which mainly compensates the topographic influence of the top-surface topography and the influence due to the varying thickness of the weathering layer. This correction is intended to compensate for the distortions caused by large-scale variations encountered in the investigated area. Thus, small-scale variations still remain as small traveltime distortions in the data which the residual static correction accounts for. These residual statics can still deteriorate the subsequent stacking process that much resulting in a stacked section of poor quality.

Residual static correction methods are of great interest to further improve not only the signal-to-noise ratio after stacking, they can also enhance the reflection event continuity. This is even of greater interest in the process of building a structural image of the subsurface. In this thesis, a conventional residual static correction method has been adapted to the Common-Reflection-Surface (CRS) stack method which provides additional information about the subsurface by means of kinematic wavefield attributes. These CRS attributes parameterize a stacking surface within a spatial aperture rather than within the common-midpoint gathers, only. With the knowledge of the CRS attributes, the offset dependency of the traveltimes associated with reflection events can be eliminated which is mandatory for the determination of residual statics.

The CRS stack method is based on the paraxial ray theory. Thus, the derivation of the second-order traveltime approximation of reflection events is reviewed before some residual static correction methods are discussed. Afterwards, the CRS-based residual static correction approach is described in details. Here, many different processing parameters are presented. Each processing parameter can have a strong influence on the results concerning the quality of the residual static corrections or concerning the economical aspects of how fast an estimate of the residual static corrections converges. Only some of these aspects are addressed in the frame of this thesis as there are too many possible combinations of the processing parameters. Nevertheless, the most promising results of two synthetic and two real data examples are presented. These examples illustrate some advantages and drawbacks of the CRS-based residual static correction approach. With the comparison to the results of a conventional method available for one real data example, the CRS-based method has emphasized its improvements due to the quality of the simulated ZO section. Furthermore, the CRS-based residual static correction method is implemented into the CRS-based reflection seismic imaging workflow which is a highly automated time-to-depth imaging workflow. The level of automatization is also an important aspect in present processing of seismic reflection data due to the dramatically increased amount of recorded data.

Zusammenfassung

Vorbemerkung

Da die vorliegende Dissertation mit Ausnahme dieser Zusammenfassung in englischer Sprache verfasst wurde, wurde auf eine Übersetzung der meisten Fachbegriffe verzichtet. Größtenteils sind die Fachbegriffe in ihrem englischen Ausdruck schon in den deutschen Sprachgebrauch der Geophysik eingegliedert worden und lassen sich auch schwer durch eine adäquate deutsche Übersetzung wiedergeben. Diese Fachbegriffe, mit Ausnahme ihrer großgeschriebenen Abkürzungen, werden im weiteren Verlauf durch eine *kursive* Schreibweise hervorgehoben. Des Weiteren wird auf die erneute Darstellung von Bildern in dieser Zusammenfassung verzichtet.

Einleitung

Heutzutage ist es den Menschen möglich, jeden Ort auf der Erdoberfläche innerhalb von Stunden zu erreichen. Außerdem hat die Menschheit Möglichkeiten gefunden, sich in die Tiefen der Meere oder die Höhen der Atmosphäre, ja sogar schon bis zu anderen Planeten zu bewegen. Dies ermöglicht zumindest theoretisch eine direkte Untersuchung von Strukturen, Materialeigenschaften und dynamischen Prozessen in diesen Bereichen. Nichtsdestotrotz ist das Innere unserer Erde immer noch ein „unentdecktes Land, das noch nie ein Mensch zuvor betreten hat". Auch wenn durch Bohrlöcher oder Stollen bereits die Erdoberfläche bis in Tiefen von bis zu 10 km vereinzelt erkundet wurde, so sind das nur Kratzer an der Oberfläche im Vergleich zum mittleren Erdradius von ca. 6378 km. Daher werden Methoden zur indirekten Untersuchung des Erdinneren notwendig.

Mithilfe von vielen verschiedenen indirekten Messmethoden ist es möglich zahlreiche Materialeigenschaften passiv oder aber auch aktiv zu bestimmen. Das Erdmagnetfeld oder die Temperaturverteilung werden meist passiv gemessen. In der Seismologie hingegen werden die Ausbreitungsgeschwindigkeiten im Untergrund aus den Messungen elastischer Wellen an der Erdoberfläche berechnet. Die elastischen Wellen werden hierbei von Erdbeben erzeugt und durchlaufen die Materialien auf dem Weg zum Empfänger und übermitteln somit einen integralen Zusammenhang der Materialeigenschaften. Aber es gibt auch aktive Messmethoden, die gerade in der Explorationsseismik häufig zum Einsatz kommen. Aktiv bedeutet hierbei, dass für die Erzeugung der elastischen Wellen zum Beispiel Detonationen oder Vibratoren also künstliche Quellen eingesetzt werden. Dabei breitet sich die erzeugte elastische Welle im Untergrund aus und wird an Diskontuinitäten reflektiert. Die Reflexionen können dann an der Erdoberfläche aufgezeichnet werden und können dann genutzt werden, um Rückschlüsse auf die Verteilung der seismischen Geschwindigkeiten zu ziehen. Meist geben diese reflexionsseismischen Aufzeichnungen auch Hinweise über die geologische Struktur im Untergrund. Mit der weiteren

Analyse der Geschwindigkeitsstrukturen im Untergrund wird in der Explorationsseismik hauptsächlich nach Erdöllagerstätten gesucht. Aber auch für die Abschätzung von Erdbebenrisiken und Vulkanausbruchsgefahren dienen die Geschwindigkeitsabbilder des Untergrundes. Im weiteren Verlauf dieser Arbeit wird daher eine Methode der Verarbeitung von reflexionsseismischen (d. h., indirekten) Messungen beschrieben, um ein interpretierbares Abbild des Untergrundes zu erhalten.

Die Erfassung seismischer Daten in der zweidimensionalen Reflexionsseismik wird hauptsächlich mit *common-shot* (CS) Anordnungen durchgeführt (siehe Abbildung 1.1(a)). Dabei wird die „Antwort" des Untergrundes auf ein seismisches Quellsignal aufgenommen. Die aufgezeichneten seismischen Spuren, die in einer CS-Sektion dargestellt werden, gehören zu einem Ereignis z. B. einer Explosion, die eine sich im Untergrund ausbreitende seismische Welle auslöst. Der *half-offset* ist die halbe und der *offset* die ganze Distanz zwischen Quell- und Empfängerposition für jedes Schuss-Empfänger-Paar. Die *midpoint*-Koordinate ist der Mittelpunkt zwischen der Quell- und Empfängerposition. Die erzeugte Welle wird an Diskontinuitäten im Untergrund refraktiert oder reflektiert. Die Teile der reflektierten oder refraktierten Welle, die die Empfänger erreichen, werden anhand der vergangenen Zeit relativ zur Auslösung der Quelle, d. h. der Laufzeit, aufgezeichnet. Die somit gewonnen Zeitreihen (auch als Spuren bezeichnet) werden in einer CS-Sektion nach aufsteigendem *offset* beziehungsweise *half-offset* sortiert. Die CS-Anordnung wird dann im Falle zweidimensionaler Messungen entlang einer geraden seismischen Linie verschoben, um viele CS-Sektionen zu erhalten, die Reflexionen von jeweils gleichen Punkten aufgezeichnet mit verschiedenen *offsets* im beleuchteten Untergrund enthalten. Für marine Messungen ist eine Gerade als seismische Linie leichter realisierbar, da z. B. das Messschiff nur den vorgegebenen Koordinaten nachfahren muss. Für auflandige Messungen ist dies nicht immer möglich. Die Topographie der Erdoberfläche oder die Infrastruktur verhindern, so dass Messungen an Land nicht immer einer geraden Linie folgen. Die zumindest meist stückweise gerade seismische Linie wird als so genannte *crooked line* bezeichnet(für detailliertere Definitionen siehe Sheriff, 2002). Sind die Abweichungen von der Ausgleichsgerade durch alle Quell- und Empfängerpositionen klein im Vergleich zu der Länge der seismischen Linie, so können die Quell- und Empfängerpositionen ohne weitere Korrekturen auf die Ausgleichsgerade projiziert werden. Somit bilden alle aufgezeichneten Sektionen den so genannten *multicoverage* Datensatz, der eine gewisse Redundanz enthält, da einige Tiefenpunkte unterhalb der zweidimensionalen reflexionsseismischen Linie mehrfach überdeckt sind.

Verschiedene Umsortierungen der aufgenommenen Spuren können vorgenommen werden, um andere Sektionen zu bilden, die einen Schritt auf dem Weg zur Interpretation der aufgenommenen Daten repräsentieren. Eine Möglichkeit kann hierbei als das Gegenstück zur CS-Sektion betrachtet werden. Anstatt Spuren von nur einer Quellposition werden Spuren einer Empfängerposition zusammengefasst zu einer *common-receiver* (CR) Sektion (siehe Abbildung 1.1(b)). Eine weitere Umsortierung ist die *common-offset* (CO) Sektion. Eine CO-Sektion beinhaltet alle Spuren mit einem bestimmten konstanten *offset*, die nach ihrer *midpoint*-Koordinate sortiert sind (siehe Abbildung 1.1(c)). Eine spezielle CO-Sektion ist die *zero-offset*-Sektion. Hier ist der *(half-) offset* Null, das heißt, dass die Positionen eines Quell-Empfänger-Paares zusammenfallen (siehe Abbildung 1.1(d)). Jedoch kann diese ZO-Anordnung bei der reflexionsseismischen Datenerfassung nicht realisiert werden, da die Quelle möglicherweise den koinzidenten Empfänger zerstört. Die ZO-Sektion muss üblicherweise von Stapelungsmethoden simuliert werden.

Die *common-midpoint* (CMP) Sektion kombiniert alle Spuren des gleichen *midpoint* sortiert nach aufsteigenden *(half-) offsets* (siehe Abbildung 1.1(e)). Die ersten fünf Darstellungen in Abbildung 1.1 verdeutlichen die Strahlenwege der meist verbreiteten Anordnungen am Beispiel eines horizontalen Reflektors mit einem Überbau konstanter Geschwindigkeit. Die CMP-Sektion wird manchmal auch als *common-depth-point* (CDP) Darstellung bezeichnet, was aber nur für den Fall von ebenen horizontalen Schichten richtig ist. Dort ist die x-Koordinate des CMP und des CDP gleich. Sobald der Reflektor aber nicht mehr horizontal ist (siehe Abbildung 1.1(f)), unterscheiden sich die x-Werte des CMP und des CDP. Daher erreichen die Strahlen für eine CMP-Konfiguration den ebenen geneigten Reflektor in einem verschmierten Bereich.

Innerhalb dieser Arbeit werden alle aufgenommenen Spuren in einem dreidimensionalen Datenraum platziert. Dieser Datenraum ist durch die folgenden Achsen bestimmt: die x_m-Achse bezeichnet die *midpoint*-Koordinate, die h-Achse steht für den *half-offset* und die t-Achse gibt die vergangene Zeit seit Auslösung der Quelle an. Die fünf in Abbildung 1.1 erwähnten Sektionstypen sind in diesem dreidimensionalem Datenraum enthalten. Abbildung 1.2 zeigt die zu CS-, CO- und CMP-Sektionen gehörigen Ebenen. Rot eingefärbte Ebenen bezeichnen CS-Sektionen. Jede CS-Sektion ($x_m - h =$ konst.) schließt mit jeder Ebene $h =$ const. oder $x_m =$ const. einen Winkel von 45 Grad ein. Grün eingefärbte Ebenen sind Beispiele für CMP-Sektionen. Innerhalb einer CMP-Sektion ist der *midpoint* konstant ($x_m =$ const.), d. h. gleich (engl. *common*) für alle enthaltenen Spuren. Die blau eingefärbten Ebenen stellen einige CO-Sektionen ($h =$ const.) dar. Hierbei ist der *half-offset* beziehungsweise *offset* konstant für alle Spuren einer CO-Sektion. Der Spezialfall einer ZO-Sektion ist die vorderste Ebene dieses Datenraumes, d. h., die Ebene $h = 0$.

Der beschriebene Datenraum wird für die *common-reflection-surface* (CRS) Stapelung (engl. *stack*) verwendet. Stapeln bedeutet hierbei, die Summation aller Amplituden einer Sektion entlang einer Laufzeitkurve oder Laufzeitfläche in den Daten. Falls die Laufzeitkurve der realen Kurve eines Reflexionsereignisses in den Daten entspricht, dann werden die kohärenten Amplituden konstruktiv aufsummiert. Das Ergebnis wird im entsprechenden Punkt der ZO-Sektion platziert. Der Nutzen ist ein höheres Signal-zu-Rauschen (engl. *signal-to-noise* (S/N)) Verhältnis, welches die Identifizierung von Reflexionsereignissen erleichtert. Das S/N-Verhältnis ist definiert als das Verhältnis der maximalen Amplitude aller Reflexionsereignisse eines Datensatzes zur *root mean square* Amplitude des Rauschens. Ein S/N-Verhältnis kleiner als eins bedeutet, dass die Signale eines Reflexionsereignisses meist nicht visuell erfasst werden können, da ihre Amplituden kleiner sind als die des Rauschens. Daher wird ein hoher S/N-Wert angestrebt. Die so genannte CMP-Stapelung ist ein Beispiel für die Summation von Amplituden in einer CMP-Sektion entlang einer Laufzeitkurve. Im Gegensatz zu der CMP-Stapelung legt die CRS-Stapelungsmethode eine Stapelfläche innerhalb des 3D Datenraumes fest. Im Falle des 2D ZO CRS-Stapelung ist die Stapelfläche durch drei kinematische Wellenfeldattribute parametrisiert, wobei diese CRS-Attribute eine physikalische Bedeutung haben und geometrisch interpretiert werden können. Das Ziel der CRS-Stapelung ist wesentlich mehr kohärente Amplituden aufzusummieren. Somit steigt das S/N-Verhältnis weiter an.

Bei auflandigen Messungen hat die Topographie der Erdoberfläche einen starken Einfluss auf die Form der Reflexionsereignisse. Für eine ebene Messoberfläche oberhalb eines horizontal geschichteten Mediums können die Reflexionsereignisse durch Hyperboloide im 3D Datenraum angenähert werden. Die Topographie der Erdoberfläche kann die Form der Reflexionsereignisflächen zu beliebig variierenden Flächen verändern. Daher wurde von Zhang (2003) die Laufzeitflächenapproximationsformel zweiter Ordnung in der CRS-Stapelungsmethode für ebene Messoberflächen generalisiert, um auch die Behandlung von komplexen Messoberflächen bei der Anwendung der CRS-Stapelmethode nutzen zu können.

Abbildung 1.3 veranschaulicht viele verschiedene Phänomene, die bei der Bearbeitung reflexionsseismischer Daten berücksichtigt werden sollten. Wie schon oben erwähnt, ist die Topographie der Erdoberfläche eines dieser Phänomene. Ein weiterer Aspekt ist das Rauschen. Unabhängig, ob das Rauschen natürlichen oder menschlichen Ursprungs ist, wird sein Einfluss deutlich durch Stapelmethoden reduziert. Die Unterdrückung von kurz- und langperiodischen Multiplen ist noch immer ein schwieriges Unterfangen. Neben all diesen Phänomenen sind Landdatensätze zusätzlich noch durch die oberste Schicht langsamer Ausbreitungsgeschwindigkeiten gestört. Hierbei hat die so genannte Verwitterungsschicht eine sehr stark variierende Geschwindigkeitsverteilung, die anhand des Streuers rechts in Abbildung 1.3 angedeutet ist. Wie der Name Verwitterungsschicht vermuten lässt, so ist diese Schicht hauptsächlich durch Verwitterung hervorgerufen. Mit einigen Annahmen, die später detaillierter erklärt werden, wird der Einfluss der Verwitterungsschicht normalerweise in zwei Schritten kompensiert. Zuerst wird der topographische Einfluss durch so genannte Feld- oder Datumsstatikkorrekturen beseitigt. Danach können noch verbliebene Statiken, die die zuvor angewendeten Korrekturen nicht kompensiert haben, aber auch zusätzlich durch die Datumsstatikkorrektur eingeführte Fehler mit Hilfe der so genannten Reststatikkorrektur behoben werden. Diese reststatische Korrektur und die Einbindung ihrer Bestimmung in die CRS-Stapelungsmethode ist das Hauptziel dieser Arbeit.

Da die Menge an aufgezeichneten Daten entlang seismischer Linie in den letzten Jahrzehnten dramatisch zugenommen hat und die Rechenleistung exponentiell angestiegen ist, ist die Nachfrage nach automatisierten Bearbeitungsschritten entsprechend angestiegen (siehe Roth, 2004). Somit wird in dieser Arbeit die CRS-basierte Reststatikkorrekturmethode vorgestellt, die inzwischen auch Bestandteil eines CRS-basierten Bearbeitungsablaufes ist. Dieser Bearbeitungsablauf ist dabei ein hochgradig automatisierter Ablauf zur Erstellung eines Tiefenabbildes des Untergrundes. Bevor aber nun die vielversprechenden Ergebnisse im Verlaufe dieser Arbeit präsentiert werden, wird der theoretische Hintergrund des neuen CRS-basierten Ansatzes zur Bestimmung reststatischer Korrekturen erklärt. Abbildung 1.4 zeigt eine kleine Vorschau auf mögliche Verbesserungen erreicht durch statische Korrekturen. Abbildung 1.4(a) ist die simulierte ZO-Sektion vor der statischen Korrektur und Abbildung 1.4(b) nach der statischen Korrektur. Deutlich sichtbar sind zum einen die verbesserte Kontinuität der Reflexionsereignisse in der Mitte der Abbildungen und zum anderen, dass die Reflexionsereignisse nach der Korrektur wesentlich flacher erscheinen, was in diesem Fall wohl auch besser der wahren Untergrundstruktur entspricht. Dieses und weitere Beispiele können bei Yilmaz (1987), Cox (1999), Yilmaz (2001a) und Yilmaz (2001b) nachgeschlagen werden.

Schlussfolgerungen

Die vorliegende Arbeit kann grob in zwei Teile gegliedert werden. Zu Anfang befasse ich mich mit den theoretischen Hintergründen der CRS-Stapelungsmethode, der statischen Korrekturen und im speziellen der reststatischen Korrekturen, wie sie in die neue CRS-basierte Reststatikkorrekturmethode eingebunden wurden. Hierzu, nehme ich Bezug auf die notwendigen Einschränkungen, um die Strahlenseismik anwenden zu können. Dabei hat die CRS-Stapelungsmethode schon öfter gezeigt, dass sie gegenüber konventionellen Methoden dank ihrer räumlichen Stapelapertur ein höheres S/N-Verhältnis erreicht. Des Weiteren gehe ich davon aus, dass die größere Anzahl an Spuren innerhalb der Stapelapertur einen positiven Einfluss auf die Zuverlässigkeit der Bestimmung reststatischer Korrekturen mit Hilfe der CRS-Stapelungsmethode hat.

Im zweiten Teil steht eher die praktische Anwendung der neuen Methode im Vordergrund. Anhand der in dieser Arbeit vorgestellten synthetischen und realen Datenbeispiele konnte gezeigt werden, dass die CRS-basierte Reststatikkorrekturmethode im Vergleich mit konventionellen Methoden zur Bestimmung der Verschiebungszeiten konkurrenzfähig ist. Hierbei ergaben sich zwar auch Verschlechterungen im Falle des ersten synthetischen Datenbeispiels, die aber durch die geeignete Wahl der Bearbeitungsparameter behoben werden können. Das zweite synthetische Datenbeispiel diente aufgrund seines geringen Verbrauchs an Rechenzeit als Datenbasis für Tests mit verschiedenen Bearbeitungs-parametersätzen. Dabei konnten Vorteile und aber auch Nachteile der bereits implementierten Variationen der Bearbeitungsschritte hervorgehoben werden. Somit hat sich ein Bearbeitungsparametersatz herauskristallisiert, der auch im weiteren Verlauf dieser Arbeit seine Anwendung findet. Aus dem Vergleich mit konventionell errechneten Reststatikkorrekturen anhand des ersten Realdatenbeispiels zeigte sich, dass die CRS-basierte Reststatikkorrekturmethode zwar nicht immer wesentliche Verbesserungen ergibt, aber zumindest gleichwertige Ergebnisse erzielt werden konnten. Dennoch kann die Anwendung der CRS-basierten Methode nach der bereits erfolgten konventionellen Korrektur immer noch Verbesserungen ergeben. Somit kann man schließlich sagen, dass die CRS-basierte Reststatikkorrekturmethode empfohlen werden kann, um Verbesserungen des Signal-zu-Rauschen Verhältnisses und der Kontinuität von Reflexionsereignissen zu erhalten. Auch sollte man im Hinterkopf behalten, dass die Methode ein Teil eines ganzen CRS-basiertem Bearbeitungsablaufes ist, der aufgrund der heutzutage immensen Datenmengen eine hohen Grad an Automatisierung besitzt. Dennoch kann damit nicht die menschliche Beurteilung der Ergebnisse ersetzt werden.

Contents

Chapter 1

Introduction

Nowadays, people can reach almost any location on earth's surface within hours. Furthermore, mankind is able to access the deep sea and even the atmosphere or other planets are in reach. Thus, a direct investigation of the structures, material properties, and dynamic processes in these environments is not always feasible but at least possible. Nevertheless, the earth's interior remains only scratched by boreholes or mining. The deepest borehole so far reached a depth of around 10 km which is relatively small compared with the average earth radius of around 6378 km. Consequently, indirect investigation methods have to be considered to gain knowledge of the earth's interior.

To fulfill the task of indirectly investigating the earth's interior, many different geophysical methods have been introduced. These indirect methods can be divided into passive measurements, e. g., of the earth's gravity and magnetic field and into active measurements, e. g., with artificial electric and electro-magnetic fields. In the framework of this thesis, other indirect methods based on the propagation of elastic energy are used. These methods can again be divided into passive and active measurements. Passive measurements are mainly used in seismology to investigate the global structure of the earth's interior by means of the propagation of elastic energy induced by earthquakes. In exploration seismics, usually active measurements are performed to gain knowledge of the structures up to a depth of around 20 km using artificial energy sources (e. g., detonations, vibrators, etc.). The aim of such active methods in exploration seismics is mainly to find hydrocarbon reservoirs, but also to estimate the risk of earthquakes or of volcanic hazards. Therefore, the knowledge of the subsurface structures is necessary. The thesis on hand describes processing methods that help to obtain an interpretable image of the subsurface out of reflection seismic data.

Seismic data acquisition in 2D reflection seismics is mainly performed with a common-source (CS) configuration (see Figure 1.1(a)) to obtain the subsurface reflection response of the seismic source signal. The recorded seismic traces plotted in a CS section or CS gather belong to one experiment, where, e. g., an explosion event initiates a seismic wave that propagates through the subsurface. The half-offset is half the distance and the offset is the entire distance between the source location and the receiver location for every source-receiver pair. The midpoint coordinate is the center between the source and the receiver location. The generated wave is refracted and reflected at discontinuities in the subsurface. Those parts of the reflected or refracted wave that emerge at the receivers are recorded with respect to the elapsed time relative to the excitation of the source, i. e., the traveltime. The recorded time series (also called traces) within a CS gather are sorted with increasing offset or half-offset, respectively. For 2D measurements, the CS configuration is then moved along a straight seismic line to obtain many CS sections that contain reflections of the same reflector points in the illuminated

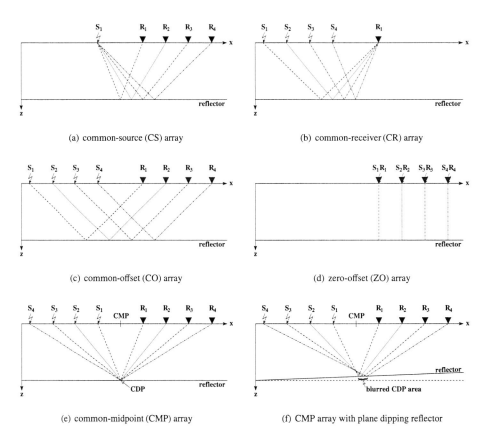

(a) common-source (CS) array

(b) common-receiver (CR) array

(c) common-offset (CO) array

(d) zero-offset (ZO) array

(e) common-midpoint (CMP) array

(f) CMP array with plane dipping reflector

Figure 1.1: Five different seismic array examples. Ray paths of five different seismic arrays that form: (a) a common-source (CS) gather, mainly used for data acquisition, (b) a common-receiver (CR) gather, (c) a common-offset (CO) gather, (d) a zero-offset (ZO) gather which is a special CO gather, and (e) a common-midpoint (CMP) gather with the common-depth-point (CDP). (f) ray paths of a CMP gather for a plane dipping reflector. Here, a CDP cannot be assigned exactly anymore but the blurred area of the depth points is taken as CDP as long as some assumptions are satisfied.

subsurface recorded with different offsets. For offshore measurements, a straight seismic line is easier to realize as, e. g., a ship has only to follow the provided coordinates. For onshore measurements, this is not always possible. The top-surface topography or the infrastructure often prevents an onshore measurement from being performed along a straight seismic line which then results in a so-called crooked line (for further definitions refer to Sheriff, 2002). If the deviations of the midpoints from the line of regression through all source and receiver locations are relatively small compared to the total length of the seismic line, the source and receiver locations can be projected to the line of regression without further corrections. Then, the entire recorded sections form a so-called multicoverage dataset and contain a certain redundancy as the data is assumed to cover depth points beneath the 2D reflection seismic line multiple times.

Several rearrangements of the recorded traces can be performed to form other sections that represent one step on the way to interpret the recorded data. One way can be regarded as the opposite of a CS gather. Here, instead of one source location one receiver location is considered which then forms the common-receiver (CR) gather (see Figure 1.1(b)). Another one is to resort the traces into a common-offset (CO) gather. A CO gather contains all traces with a certain constant offset that are sorted by their midpoint coordinate (see Figure 1.1(c)). A special CO gather is the zero-offset (ZO) gather. Here, the (half-) offset is zero, i.e., the source and receiver locations for a ZO source-receiver pair coincide (see Figure 1.1(d)). However, this ZO configuration cannot be realized for reflection seismic acquisitions as the source would possibly destroy the coincident receiver. Thus, the ZO section usually has to be simulated by stacking methods.

The common-midpoint (CMP) gather combines all traces with the same midpoint and sorts them with increasing (half-) offsets (see Figure 1.1(e)). The first five parts of Figure 1.1 illustrate the ray paths belonging to the most commonly used configurations for the example of a horizontal reflector with a constant velocity overburden. The CMP gather sometimes is also called common-depth-point (CDP) gather which is only correct for the case of plane horizontal layers. There, the x-coordinate of the CMP and the CDP are the same. But as soon as the reflector is not horizontal anymore (see Figure 1.1(f)), the x-coordinate of the CMP and the CDP differ. Furthermore, an exact CDP cannot be assigned anymore. If the reflector dip remains small, the area containing the reflection points along the reflector can still be regarded as CDP. Thus, the rays for a CMP configuration emerge at the plane dipping reflector in a blurred CDP area.

For this thesis, all recorded data are located in a three dimensional data volume with the following axes: the x_m-axis denotes the midpoint coordinate, the h-axis stands for the half-offset, and the t-axis is the time elapsed since excitation of the source. The five different gather types mentioned in Figure 1.1 are contained in this 3D data volume. Figure 1.2 shows the planes belonging to CS, CO, and CMP gathers. Red colored planes denote CS gathers. Every CS gather ($x_m - h = $ const.) builds with every plane $h = $ const. or $x_m = $ const. an angle of $45°$ in the x_m-h plane. The CR gathers are perpendicular planes to the CS gathers. Green colored planes are examples for CMP gathers. Within one CMP gather, the midpoint is constant ($x_m = $ const.), i.e., common for all contained traces. The blue colored planes depict some CO gathers ($h = $ const.). There, the half-offset is constant for all traces within one CO gather. The special case of a ZO gather is the front plane of this data volume, i.e., the plane $h = 0$.

The described data volume is the input for the common-reflection-surface (CRS) stack. Stacking means to sum up all amplitudes in a section along a traveltime curve or surface in the data volume. If the traveltime curve or surface coincides with the true reflection event the coherent amplitudes are summed up constructively. The result is assigned to the corresponding point of the ZO section. The benefit is a higher signal-to-noise (S/N) ratio that makes it easier to identify reflection events. The S/N ratio is defined as the quotient of the maximum amplitude of all reflection events within one dataset over the root-mean-square amplitude of the noise. A S/N ratio smaller than one means that the signals of the reflection events mostly cannot be distinguished visually from the noise because their amplitudes are smaller than those of the noise. Thus, a high S/N value is desired. The so-called CMP stack is one example for summing up amplitudes in CMP gathers along a traveltime curve. In contrast to the CMP stack, the CRS stack method determines a stacking surface within the 3D data volume. In case of the 2D ZO CRS stack, this surface is parameterized by three kinematic wavefield attributes that have a physical meaning and can be geometrically interpreted. The aim of this CRS stacking

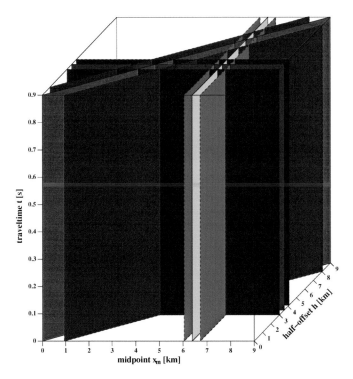

Figure 1.2: 3D multicoverage data volume of a 2D measurement. The coordinates along the axes of this 3D data volume are the midpoint coordinate x_m, the half-offset h, and the traveltime t. The colored planes indicate subsets commonly used for analyzing and interpreting the recorded data. The red planes stand for common-source gathers and represent the most common acquisition geometry. The blue planes are common-offset gathers and the green planes show examples of common-midpoint gathers. Common-receiver gathers are not displayed as they are simply perpendicular to the common-source gathers.

surface is to sum up even more coherent amplitudes. Thus, the S/N ratio increases even more.

In case of onshore measurements, the top-surface topography along the 2D seismic line may strongly influence the shape of the reflection events. For a planar measurement surface above a horizontally stratified medium, the reflection event can be approximated by hyperboloids in the 3D data volume. A rough top-surface topography can change the shape of the reflection event surfaces to arbitrarily varying surfaces. Therefore, the second-order traveltime approximation formula of the CRS stack method for planar measurement surfaces has been generalized by Zhang (2003) to consider the top-surface topography directly within the application of the CRS stack method.

Figure 1.3 illustrates many different phenomena that have to be considered in the processing of seismic reflection data. As already mentioned, the top-surface topography is illustrated. Another aspect is the noise. Irrespectively whether natural or human-made noise, its influence can be significantly reduced by the mentioned stacking methods. The suppression of short period and/or long period

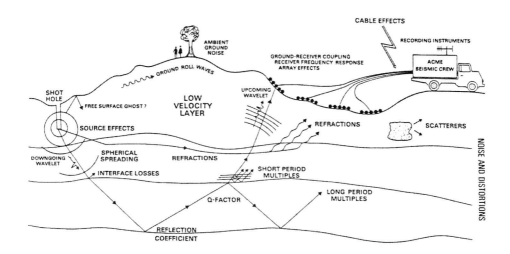

Figure 1.3: Phenomena causing degradation of seismic waves. Several phenomena concerning the subsurface structure like (natural and artificial) noise, different kinds of multiples and of course also the low-velocity layer are visualized. Furthermore, phenomena due to the acquisition are indicated, e. g., cable effects within this figure (taken from Reynolds, 1997).

multiples is still a difficult task. Besides these phenomena, land datasets are further distorted by a low-velocity layer (LVL) as the uppermost layer which is usually strongly inhomogeneous as indicated by a scatterer shown on the right hand side in Figure 1.3. This LVL is also called weathering layer as it is mainly induced by weathering. With several assumptions explained in more details later on the influence of this LVL is usually eliminated in two steps. Firstly, the topographic influence is removed by so-called field static or datum static corrections. Secondly, the remains which the previous corrections did not account for and also some errors introduced by datum static corrections can be compensated by so-called residual static corrections. These residual static corrections and their estimation within the CRS stack method are the main target of this thesis.

As the amount of data recorded along seismic lines has grown dramatically and the CPU power has accelerated exponentially, the demand for automated processing methods has increased (see Roth, 2004). Thus, in this thesis, a CRS-based residual static correction method is presented which is already integrated into the CRS-based imaging workflow, i. e., a highly automated time-to-depth domain imaging workflow. Nevertheless, before I can show the promising results of the new CRS-based approach for residual static correction, I review the theoretical background of the CRS stack method and residual static correction methods. Nevertheless, to give a little motivation for the presented new approach, I show the comparison of two ZO sections from a real dataset to emphasize the improvements obtained by static corrections. For this example and others, please refer to Yilmaz (1987), Cox (1999), and Yilmaz (2001a) and Yilmaz (2001b). Figure 1.4(a) displays the ZO section directly simulated from the real dataset. It is obvious that the applied stacking method was not able to successfully sum up coherent reflection events in the middle of this section. Thus, static corrections have been applied and Figure 1.4(b) shows the resulting ZO section. Two main differences can be observed, on the one hand, the reflection events are now visible throughout the entire simulated ZO section and, on

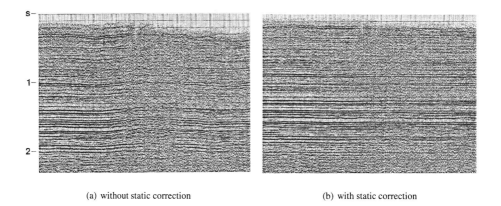

(a) without static correction   (b) with static correction

Figure 1.4: Example of a simulated ZO section before and after static correction. (a) shows a real data example before static corrections have been applied. In the middle hardly any reflection events can be recognized. Furthermore, the reflection events are not flat. (b) displays the same real data set after static corrections have been applied. It is obvious that the reflection event continuity has been improved. Also, the reflection events are almost perfectly flat within this ZO section (taken from Yilmaz, 1987).

the other hand, the reflection events have been flattened and are now closer to the expected geologic structure in the subsurface.

1.1   Structure of the thesis

Throughout this thesis, the subsequent chapters can be separated into two main parts: a theoretical part (Chapters 2-4) and a practical part (Chapters 5 and 6). In the theoretical part, I briefly describe the basic ideas that enter into the considerations made in the scope of this thesis. Thus, I will begin with an introduction to the fundamentals of ray theory in Chapter 2. There, the elastodynamic wave equation is derived and one of its solutions is discussed. Furthermore, paraxial ray theory is used to obtain a second-order traveltime approximation for reflection events.

In Chapter 3, I present a derivation of an exact formula for the Common-Reflection-Point (CRP) trajectory in case of unconverted waves that traverse a homogeneous, isotropic medium and are measured at a planar surface. With the concepts of geometrical optics, the CRP trajectory for inhomogeneous, isotropic overburden and a curved reflector segment (or CRS) is approximated for the kinematic reflection ZO response again at a planar surface. This geometrical derivation should give a brief insight into the geometrical interpretation of the CRS attributes parameterizing the reflection response in the time domain. Furthermore, the traveltime formula derived in Chapter 2 is simplified for the 2D ZO case. However in contrast to the geometrical derivation, this second traveltime formula approximates the reflection response obtained along an arbitrarily rough top-surface topography. One of the most commonly used conventional stacking methods, i. e., the NMO/DMO/stack method, is briefly compared with the CRS stack method. Finally, the limits of applicability of the CRS stacking operator is addressed by describing the CRS aperture.

After the theoretical background of the CRS stack is described, the problem of residual statics is discussed. Chapter 4 gives a brief introduction to the problem of statics introduced by the weathering layer which is mainly encountered in onshore datasets. The assumptions necessary to successfully remove the influence of the LVL are described. The influence is usually removed in two steps: firstly, the field static corrections are briefly discussed. These field static corrections are used to eliminate mainly the topographic influence of the weathering layer. Nevertheless, some residual statics might remain in the dataset. To further reduce such remainders, the implementational differences of some of the most commonly used conventional methods are compared with the new CRS-based residual static correction approach. I introduce this new approach based on the CRS wavefield attributes and the simulated ZO section. The estimation of the time shifts and other implementational aspects are discussed in more details.

Two synthetic datasets are used in Chapter 5 to demonstrate the potential of the CRS-based residual static correction approach. Therefore, synthetic data example A is used to obtain a first test result to see whether artificially added random but surface-consistent time shifts can be recorrected by the new approach. Synthetic data example B is a small and simple subset of a real dataset which has also been distorted by artificial time shifts to simulate residual statics. As the dataset is small, it is well suited for the testing of different parameter sets of the new CRS-based approach. Thus, the results of several iterations applied to both datasets are presented and discussed.

Chapter 6 is also divided in two parts as each part presents the results of a different real data example. Furthermore, real data example A consists of two 2D recordings of two parallel seismic lines measured in the Upper Rhine Graben. As the contractor has additionally provided the results of a conventional residual static correction method, the CRS-based residual static correction method is compared with the conventional results. Real data example B is the first example where the rough top-surface topography has to be considered during the CRS processing and the CRS-based residual static correction approach. Unfortunately, no conventional residual static corrections are available for this dataset. Thus, only the obtained results are presented and the improvements of the stack results are discussed.

I summarize the experiences and results of the previous chapters in Chapter 7. Furthermore, a brief outlook for future developments is given.

Appendix A lists most of the notation used in this thesis. Furthermore, a small list of the used abbreviations is contained. In Appendix B, a brief discussion of important properties of the propagator matrix which are used in the theoretical derivation of the traveltime formula utilized in the CRS stack method is given. Refraction seismics are commonly used to obtain the depth of the base of weathering and the velocity of the LVL. Furthermore, methods like the Plus-Minus method by Hagedoorn (1959) are used to estimate time shifts from refraction seismic measurements that are usually also contained in reflection seismic measurements. However, to explain all the aspects concerning the estimation of refraction statics is way too far for the frame of this thesis. Thus, Appendix C describes the simplest case of refraction seismics, i. e., a horizontal plane reflector with constant-velocity layers. The different methods for estimating time shifts based on refraction seismics can be read in many other books commonly used in seismics.

Chapter 2

Ray theory fundamentals

Seismic body wave propagation in complex, laterally varying 3D layered structures can be described in several ways. On the one hand, there are methods based on direct solutions of the elastodynamic wave equation, e. g., finite-difference (FD), Kelly et al. (1976), or finite-element (FE), Strang and Fix (1973), methods to mention the most common ones. These methods provide a solution for the wavefield on a grid as the medium is discretized. On the other hand, also high-frequency approximations asymptotically satisfy the elastodynamic wave equation adequately in many cases. Ray theory belongs to these high-frequency asymptotic methods. All the mentioned methods are suitable to solve different kinds of problems. If the investigated medium is strongly heterogeneous, i. e., the physical properties vary on a length scale equal to or smaller than the dominant wavelength of the propagating signal, FD schemes are a good choice. But if the physical properties vary smoothly within an inhomogeneous, isotropic, layered medium, ray theory is more advantageous if not the entire wavefield has to be considered. Here, the high-frequency approximation is suited to solve the elastodynamic wave equation in an effective and usually sufficiently accurate way. Under the assumption of smoothness, ray theory can be used to compute rays, traveltimes, and wavefronts (the kinematic aspects of the wave propagation process) as well as amplitudes (one dynamic aspect of the wave propagation process) depending on the selected ray code. In contrast to FD schemes, where the appearance of multiples is inevitable, multiples can be omitted in ray theory by not taking their ray code into account. Furthermore, methods based on ray theory do not have to overcome problems like grid dispersion and reflections from model boundaries. Nevertheless, ray theory may run into problems at, e. g., caustics.

In this chapter, I will present a short summary of the basic ideas and formulae underlying ray theory. For further description beyond this thesis, the reader is recommended to refer to Aki and Richards (1980), Popov (2002), and especially Červený (2001) for a profound study on ray theory. For most of the used notations and symbols, please refer to Appendix A.

2.1   The elastodynamic wave equation

Continuum mechanics is used to describe media whose properties and all physical parameters can be treated as continuous and sufficiently smooth functions of space and time. Besides the direct vicinity of the seismic source, the earth can be regarded as an elastic continuum for small deformations. In this case, the elastodynamic wave equation is used to mathematically describe wave propagation in such media. The discrete structure of a medium is not considered here.

Assume that a small particle within a solid body has a volume V and a boundary surface S. Inserting this assumption into Newton's second law of mechanics[1] yields

$$\frac{d}{dt} \iiint_V \rho \frac{d\vec{u}}{dt} \, dV = \iiint_V \vec{f} \, dV + \oiint_S T(\vec{n}) \, dS \,, \tag{2.1}$$

where the velocity is given by the temporal derivative of the displacement vector $\vec{u} = \vec{u}(\vec{x}, t)$ at location $\vec{x} = (x_1, x_2, x_3)^T$ in Cartesian coordinates and time t. $\rho = \rho(\vec{x})$ is the density distribution within the solid body. $\vec{f}$ denotes the density of the external forces and $T(\vec{n})$ describes the traction acting through the boundary surface S depending on its normal vector $\vec{n}$. The total derivative of an arbitrary function $a = a(\vec{x}, t)$ is expressed by

$$\frac{d}{dt} a(\vec{x}, t) = \frac{\partial}{\partial t} a(\vec{x}, t) + \left(\frac{d\vec{x}}{dt} \cdot \vec{\nabla} \right) a(\vec{x}, t) \,. \tag{2.2}$$

This expression accounts for the movement of the whole media which is important in fluid mechanics. Nevertheless, I assume that the solid body itself does not move, i. e., $\frac{d\vec{x}}{dt} = 0$ yields that the second term on the right side of equation (2.2) vanishes. Thus, the total derivatives with respect to time t can be replaced by their partial derivatives. Furthermore, as the density ρ is assumed constant in time t, the term on the left side of equation (2.1) simplifies and the equation of motion (2.1) reads

$$\iiint_V \rho \frac{\partial^2 \vec{u}}{\partial t^2} \, dV = \iiint_V \vec{f} \, dV + \oiint_S T(\vec{n}) \, dS \,. \tag{2.3}$$

The traction $T(\vec{n})$ can be expressed by

$$T(\vec{n}) = \underline{\tau} \cdot \vec{n} \,. \tag{2.4}$$

Furthermore, with the theorem of Gauß for integrals:

$$\oiint_S \underline{\tau} \cdot \vec{n} \, dS = \iiint_V \underline{\tau} \, dV \tag{2.5}$$

and if all functions are several times continuously differentiable, then equation (2.3) can be rewritten and is given in differential form:

$$\rho \frac{\partial^2 u_j}{\partial t^2} = f_j + \frac{\partial \tau_{ij}}{\partial x_i} \qquad \text{with} \qquad i, j = 1, 2, 3 \,. \tag{2.6}$$

Please remember that a repeated index implies summation with respect to this index (Einstein convention). The tensor $\tau_{ij} = \tau_{ij}(\vec{x}, t)$ is the so-called stress tensor, a symmetrical tensor of second order ($\tau_{ij} = \tau_{ji}$). It describes the stress condition at any location $\vec{x}$ in the medium. f_j are referred to as the source terms as they stand for the Cartesian components of the body forces (forces per volume). The set of three equations (2.6) is also known under the name "equation of motion" for a continuum. Equations (2.6) relate the density-weighted acceleration to body forces and stress gradients in the medium. The equation of motion is the most fundamental equation underlying the theory of seismics

---

[1] Newton's second law of mechanics: The acceleration $\vec{a}$ of an object as produced by a net force (vector sum of all forces acting upon an object) is directly proportional to the magnitude of the net force $\vec{F}_{net}$, in the same direction as the net force, and inversely proportional to the mass m of the object. $\vec{F}_{net} = \frac{d(m\vec{v})}{dt}$ or simply (for $m = $ const): $\vec{F}_{net} = m\vec{a} = m\frac{d\vec{v}}{dt} = m\frac{d^2\vec{x}}{dt^2}$.

and seismology, as it represents the relation of forces in the medium to measurable displacements. For a more detailed derivation of the equation of motion, please refer to Aki and Richards (1980).

With the further assumptions of small deformations and a linear anisotropic perfectly elastic solid, the generalized Hooke's law is given by

$$\tau_{ij} = c_{ijkl} \, e_{kl} \qquad i, j, k, l = 1, 2, 3 \,. \tag{2.7}$$

$e_{kl} = e_{kl}(\vec{x}, t)$ is the strain tensor of second order, which is considered symmetric like the stress tensor ($e_{kl} = e_{lk}$). The strain tensor describes the spatial gradients of the displacement field. c_{ijkl} denotes the so-called stiffness tensor or elastic tensor (i. e., tensor of elastic parameters). In general, c_{ijkl} has $3 \times 3 \times 3 \times 3 = 81$ components. These components, however, satisfy the following symmetry relations:

$$c_{ijkl} = c_{jikl} = c_{ijlk} = c_{klij} \,, \tag{2.8}$$

which reduce the number of independent components to 21 while the stress and strain tensors have six instead of nine independent components due to their symmetry. As the stiffness tensor stands for a proportionality factor in the stress-strain relationship (2.7), its components are known as the elastic moduli and define the elastic material properties of the medium.

If strains and displacement derivatives remain small, the nine elements of the strain tensor can be represented by

$$e_{kl} = \frac{1}{2} \left(\frac{\partial u_k}{\partial x_l} + \frac{\partial u_l}{\partial x_k} \right) . \tag{2.9}$$

Inserting the linear dependence of the strain components on the derivatives of displacement components (2.9) into equation (2.7) and differentiating with respect to x_i leads to

$$\frac{\partial \tau_{ij}}{\partial x_i} = \frac{1}{2} \frac{\partial}{\partial x_i} \left(c_{ijkl} \frac{\partial u_k}{\partial x_l} + c_{ijkl} \frac{\partial u_l}{\partial x_k} \right) . \tag{2.10}$$

The symmetry property $c_{ijkl} = c_{ijlk}$ of the stiffness tensor simplifies equation (2.10) as it is possible to interchange the indices k and l in the last term because of the summation according to these indices. Inserting the result into equation (2.7) yields

$$\rho \frac{\partial^2 u_j}{\partial t^2} - \frac{\partial}{\partial x_i} \left(c_{ijkl} \frac{\partial u_k}{\partial x_l} \right) = f_j \,. \tag{2.11}$$

The last set of equations is a hyperbolic system of three partial differential equations of second order which is called the elastodynamic wave equation for inhomogeneous anisotropic perfectly elastic media. For inhomogeneous media, this system of equations (2.11) has, in general, no analytical solution.

In the context of seismic ray theory, anisotropy is defined as the variation of seismic velocity depending on either the direction of propagation (for P- or S-waves) or the direction of polarization (for S-waves). Fortunately, many materials of the earth can be regarded as a medium whose elastic properties are independent of direction or polarization of propagating waves under consideration. In this isotropic case, the components of the elastic tensor can be expressed in terms of two independent elastic moduli λ and μ, namely

$$c_{ijkl} = \lambda \delta_{ij} \delta_{kl} + \mu \left(\delta_{ik} \delta_{jl} + \delta_{il} \delta_{jk} \right) , \tag{2.12}$$

11

where δ_{ij} is the Kronecker symbol,

$$\delta_{ij} = 1 \quad \text{for} \quad i = j, \qquad \delta_{ij} = 0 \quad \text{for} \quad i \neq j, \tag{2.13}$$

λ and μ are called Lamé parameters. The significance of the shear modulus, or rigidity, μ is readily apparent as a measure of a mediums resistance to shear stress. By inserting the elastic tensor (2.12) into equation (2.11) and expressing it in its vectorial form yields

$$\rho \frac{\partial^2 \vec{u}}{\partial t^2} = \vec{f} + (\lambda + \mu) \vec{\nabla} \left(\vec{\nabla} \cdot \vec{u} \right) + \mu \Delta \vec{u} + \vec{\nabla} \lambda \left(\vec{\nabla} \cdot \vec{u} \right) + \vec{\nabla} \mu \times \left(\vec{\nabla} \times \vec{u} \right) + 2 \left(\vec{\nabla} \mu \cdot \vec{\nabla} \right) \vec{u}, \tag{2.14}$$

where Δ is the Laplace operator and stands for $\vec{\nabla} \cdot \vec{\nabla} = \frac{\partial^2}{\partial x_1^2} + \frac{\partial^2}{\partial x_2^2} + \frac{\partial^2}{\partial x_3^2}$ and $\vec{\nabla} = \left(\frac{\partial}{\partial x_1}, \frac{\partial}{\partial x_2}, \frac{\partial}{\partial x_3} \right)^T$ is the Nabla operator. The last equation is the elastodynamic wave equation for inhomogeneous isotropic perfectly elastic media.

2.1.1 The acoustic wave equation

Media like water or, in a more general sense, fluids expose no significant resistance against shear stress. Thus, μ is assumed to be equal zero for fluids. These media are also called acoustic media. In seismic prospecting for hydrocarbons, the elastic medium is often approximated by an acoustic medium. Additionally, if the source term is neglected, the elastodynamic wave equation (2.14) further reduces to

$$\rho \frac{\partial^2 \vec{u}}{\partial t^2} = \vec{\nabla} \left(\lambda \vec{\nabla} \cdot \vec{u} \right) . \tag{2.15}$$

Instead of working with the displacement vector $\vec{u}$ in fluids, it is more usual to work with pressure $p = p(\vec{x}, t)$ defined by

$$p = -\lambda \vec{\nabla} \cdot \vec{u} . \tag{2.16}$$

Incorporating the pressure into equation (2.15) yields

$$\vec{\nabla} \left(\frac{1}{\rho} \vec{\nabla} p \right) = \frac{1}{\lambda} \frac{\partial^2 p}{\partial t^2} \tag{2.17}$$

which is called the acoustic wave equation. In case of a constant density media, the acoustic wave equation simplifies to

$$\frac{1}{c^2} \frac{\partial^2 p}{\partial t^2} - \Delta p = 0, \tag{2.18}$$

where $c = \sqrt{\lambda/\rho}$ is the spatially varying acoustic wave velocity.

2.1.2 Fourier transform

In the wave theory, the Fourier transformation pair is a very important tool. It connects the time domain with the frequency domain and vice versa. In the time domain, all the signals used in geophysics can be chosen to be causal. Nevertheless, some processing steps can change the signal into a zero-phase wavelet. Therefore, I will express the Fourier transform and its inverse transform in its general form.

For the definition of the Fourier transformation pair, an arbitrary temporal signal $f(t)$ is assumed which results in the following convention for forward and inverse Fourier transform:

$$F(\omega) = \int_{-\infty}^{\infty} f(t)\, e^{-i\omega t} dt, \tag{2.19a}$$

$$f(t) = \frac{1}{2\pi} \int_{-\infty}^{\infty} F(\omega)\, e^{i\omega t} d\omega. \tag{2.19b}$$

In case of causal signals, the definition (2.19a) of the forward Fourier transform reduces to the so-called causal Fourier transform where the integral has to be evaluated for positive times only. With the general Fourier transformation pair above, the elastodynamic wave equation (2.14) can be transformed to the frequency domain and reads

$$-\omega^2 \rho \vec{U} = \vec{F} + (\lambda + \mu)\, \vec{\nabla}\left(\vec{\nabla} \cdot \vec{U}\right) + \mu \Delta \vec{U} + \vec{\nabla}\lambda \left(\vec{\nabla} \cdot \vec{U}\right) + \vec{\nabla}\mu \times \left(\vec{\nabla} \times \vec{U}\right) + 2\left(\vec{\nabla}\mu \cdot \vec{\nabla}\right)\vec{U}, \tag{2.20}$$

where $\vec{U} = \vec{U}(\vec{x}, \omega)$ is the Fourier transform of the particle displacement vector $\vec{u}(\vec{x}, t)$ and $\vec{F} = \vec{F}(\vec{x}, \omega)$ the Fourier transform of the source term $\vec{f}(\vec{x}, t)$. In case of $\mu = 0$, the acoustic wave equation in its more general form with a variable density ρ reads

$$\vec{\nabla}\left(\frac{1}{\rho}\vec{\nabla}P\right) = -\frac{\omega^2}{\lambda}P, \tag{2.21}$$

where $P = P(\vec{x}, \omega)$ is the Fourier transform into the frequency domain of the scalar pressure field $p(\vec{x}, t)$ in the medium. For constant density, the acoustic wave equation in the frequency domain is given by

$$\left(\Delta + \frac{\omega^2}{c^2}\right)P = 0 \tag{2.22}$$

which is the so-called Helmholtz equation.

2.1.3 Eikonal and transport equation (acoustic case)

The next step is to consider a time harmonic wavefield

$$P(\vec{x}, \omega) = \tilde{U}(\vec{x})\, e^{-i\omega t}, \tag{2.23}$$

where ω denotes the circular frequency and $\tilde{U}(\vec{x}) = \tilde{U}$ is the scalar displacement which is independent of ω. Inserting this ansatz into the Helmholtz equation (2.22) in the frequency domain leads to

$$\left(\Delta + \frac{\omega^2}{c^2}\right)\tilde{U} = 0 \tag{2.24}$$

which is a so-called reduced wave equation for $\tilde{U}$ in the frequency domain. Here, a constant density is assumed but a variable velocity distribution $c = c(\vec{x})$. In the following, a solution in the form of

$$\tilde{U} = A(\vec{x})\, e^{i\omega \tau} \tag{2.25}$$

13

is searched for, where $\tau = \tau(\vec{x})$ is called eikonal and $A(\vec{x})$ amplitude. Please note the difference between:

$$\begin{aligned} \text{stress tensor} &\quad : \quad \tau_{ij} = \tau_{ij}(\vec{x}, t) \text{ and} \\ \text{eikonal} &\quad : \quad \tau = \tau(\vec{x}) \, . \end{aligned}$$

Then, inserting equation (2.25) into the Helmholtz equation (2.22) and using the vectorial identity $\vec{\nabla} \cdot (a\vec{b}) = \vec{b} \cdot \vec{\nabla} a + a\vec{\nabla} \cdot \vec{b}$ results in

$$e^{i\omega\tau} \left\{ \omega^2 \left(\frac{1}{c^2} - \left(\vec{\nabla}\tau\right)^2 \right) A + i\omega \left(2\vec{\nabla}\tau \cdot \vec{\nabla}A + \Delta\tau A \right) + \Delta A \right\} = 0 \tag{2.26}$$

The last equation must be valid for all frequencies which implies that each coefficient depending on ω has to vanish independently. For some simple cases like, e. g., plane waves in homogeneous media or spherical waves, there exist analytical solutions for equation (2.26) (see Červený, 2001). However in general, with only two quantities (A and τ) to be determined, this can only be achieved for order two and one of ω, i. e., for the coefficients of ω^2 and ω^1. This leads immediately to

$$\left(\vec{\nabla}\tau\right)^2 \;=\; \frac{1}{c^2} \qquad \text{(Eikonal equation)}, \tag{2.27}$$

$$2\vec{\nabla}\tau \cdot \vec{\nabla}A + \Delta\tau A \;=\; 0 \qquad \text{(Transport equation)}. \tag{2.28}$$

In the high-frequency approximation, i. e., the circular frequency ω is supposed to be sufficiently high[2], the ω^0 term is assumed to be negligible compared to the higher order terms. The exception to the rule are caustics, cusps and at points where waves are critically reflected and generate head waves. There, the ω^0 term does not remain negligible and, thus, ray theory breaks down at these locations. If nevertheless the ω^0 term increases along a ray, an infinite series for $\tilde{U}$ is introduced in order to overcome this problem. This ansatz is usually called the ray series ansatz

$$\tilde{U} = \sum_{n=0}^{\infty} \frac{A_n(\vec{x})}{(-i\omega)^n} e^{i\omega\tau} \, . \tag{2.29}$$

Substituting the ray series ansatz into the reduced wave equation (2.24) results once again in the same eikonal equation (2.27). However, a recurrent set of transport equations is obtained instead of only one transport equation

$$2\vec{\nabla}\tau \cdot \vec{\nabla}A_{n+1} + \Delta\tau A_{n+1} = \Delta A_n \, , \qquad n = -1, 0, 1, \ldots \, , \qquad A_{-1} = 0 \, . \tag{2.30}$$

This is an asymptotic series which has not to be convergent in general. For $n = -1$, the same relationship as observed in equation (2.28) for the zero-order amplitude term is obtained. Thus, most scientists restrict their work to zero-order ray theory to avoid the handling of higher order terms as this can be more difficult due to the convergence criteria.

To again generalize the approximate high-frequency equations (2.27) and (2.28), a variable density $\rho = \rho(\vec{x})$ is reintroduced. The formulae can be written in a similar way and the differences are only formal (see Červený, 2001).

---

[2] Here, circular frequency sufficiently high means: the wavelength is small compared with spatial dimensions, e. g., reflector radius of curvature or reflector thickness in layered media.

2.1.4 The elastic wave equation

Compared to the acoustic wave equation, the elastic wave equation is not constrained according to $\mu = 0$ but μ can vary depending on the location, i.e., $\mu = \mu(\vec{x})$.

In the high frequency regime, high frequency means that the wavelength ($1/\omega$) associated with this frequency should be small compared to the typical natural length scales associated with the medium. If this assumption holds true, an asymptotic solution for $\vec{U} = \vec{U}(\vec{x}, t)$ of the elastodynamic wave equation (2.20) can be obtained by the ray series ansatz

$$\vec{U} \approx \sum_{n=0}^{\infty} \frac{\tilde{\vec{U}}}{(i\omega)^n} e^{-i\omega\tau} ,\tag{2.31}$$

where the displacement vector $\tilde{\vec{U}} = \tilde{\vec{U}}(\vec{x})$ only depends on the location $\vec{x}$. The traveltime from a source to the location $\vec{x}$ is denoted with $\tau = \tau(\vec{x})$. As mentioned for the acoustic case, only the zero-order term of the ray series ansatz is used in most practical applications in seismology and seismics. Thus, the zero-order ray theory makes use of the $n = 0$ term of equation (2.31)

$$\vec{U} = \tilde{\vec{U}} e^{-i\omega\tau} .\tag{2.32}$$

Inserting the zero-order ray theory ansatz (2.32) into the elastodynamic wave equation (2.20) for inhomogeneous isotropic perfectly elastic media without any sources, i.e., $\vec{F} = 0$ and setting each coefficient depending on the order of ω equal zero, yields the following equations sorted by the order of ω:

$$\text{order} \quad : \quad \text{coefficient} \overset{!}{=} 0$$

$$\omega^2 \quad : \quad -\rho\tilde{\vec{U}} + (\lambda + \mu)\left(\vec{\nabla}\tau \cdot \tilde{\vec{U}}\right) + \mu\left(\vec{\nabla}\tau\right)^2 \tilde{\vec{U}} = 0 ,\tag{2.33}$$

$$\omega^1 \quad : \quad (\lambda + \mu)\left[\left(\vec{\nabla} \cdot \tilde{\vec{U}}\right)\vec{\nabla}\tau + \vec{\nabla}\left(\tilde{\vec{U}} \cdot \vec{\nabla}\tau\right)\right] + \mu\left[2\left(\vec{\nabla}\tau \cdot \vec{\nabla}\right)\tilde{\vec{U}} + (\Delta\tau)\tilde{\vec{U}}\right]$$
$$+ \left(\tilde{\vec{U}} \cdot \vec{\nabla}\tau\right)\vec{\nabla}\lambda + \left(\vec{\nabla}\mu \cdot \vec{\nabla}\tau\right)\tilde{\vec{U}} + \left(\vec{\nabla}\mu \cdot \tilde{\vec{U}}\right)\vec{\nabla}\tau = 0 ,\tag{2.34}$$

$$\omega^0 \quad : \quad (\lambda + \mu)\Delta\tilde{\vec{U}} + \mu\Delta\tilde{\vec{U}} + \left(\vec{\nabla} \cdot \tilde{\vec{U}}\right)\vec{\nabla}\lambda + 2\left(\vec{\nabla}\mu \cdot \vec{\nabla}\right)\tilde{\vec{U}} + \vec{\nabla}\mu \times \left(\vec{\nabla} \times \tilde{\vec{U}}\right) = 0 .\tag{2.35}$$

Ansatz (2.32) has to be a solution of equation (2.20) for each frequency ω. Thus, each coefficient depending on the order of ω has to vanish separately as indicated in equations (2.33) to (2.35). Hereby, the ω^2 coefficients (equation (2.33)) lead to the so-called eikonal equation. The eikonal equation governs the kinematic aspects of wave propagation and does not depend on the circular frequency ω. Setting the coefficients of ω^1 equal zero (equation (2.34)) yields the transport equation which governs the dynamic aspects of wave propagation. Equation (2.35) can only be fulfilled for certain simple cases[3]. However, this condition can provide a criterion for the validity of the zero-order ray series approximation (Červený, 2001).

2.1.5 The eikonal equation (elastic case)

Introducing the 3×3 matrix $\mathbf{\Gamma}$ and the identity matrix $\mathbf{I}$ makes it possible to reformulate the eikonal equation in such a way that an eigenvector equation has to be solved which reads

$$\left(\mathbf{\Gamma} - \mathbf{I}\right)\tilde{\vec{U}} = 0 \quad \text{with} \quad \mathbf{I} = \left(\delta_{ij}\right) \quad \text{for} \quad i, j = 1, 2, 3 .\tag{2.36}$$

---

[3]Some cases are, e.g., simple plane or spherical waves, but also $\tilde{\vec{U}} = 0$.

Hereby, the components of the so-called Christoffel matrix $\underline{\Gamma}$ in the special case of isotropy are defined as

$$\Gamma_{ij} = \frac{\lambda + \mu}{\rho} \frac{\partial \tau}{\partial x_i} \frac{\partial \tau}{\partial x_j} + \frac{\mu}{\rho} \delta_{ij} \left(\vec{\nabla}\tau\right)^2 . \tag{2.37}$$

Now, the eigenvector problem can be solved for its eigenvalues E_i. The necessary condition for the existence of a non-trivial solution to equation (2.36) is

$$\det\left(\underline{\Gamma} - \underline{I}\right) = 0 . \tag{2.38}$$

This results in

$$E_1 = \frac{\lambda + 2\mu}{\rho} \left(\vec{\nabla}\tau\right)^2 \quad \text{and} \quad E_2 = E_3 = \frac{\mu}{\rho} \left(\vec{\nabla}\tau\right)^2 . \tag{2.39}$$

The eigenvalues correspond to the two different observable wave-modes in elastic isotropic media. E_1 represents the longitudinal polarized wave, also known as the primary wave or shortly P-wave. E_2 and E_3 are the two mutual perpendicular transversal polarized waves, also known as secondary waves or simply S-waves, which are actually degenerated in the isotropic case.

Reinserting the eigenvalues into equation (2.36), yields that each eigenvalue has to be equal to one. This provides the eikonal equation for P- and S-waves in its commonly used form

$$\left(\vec{\nabla}\tau\right)^2 = \frac{1}{\alpha^2} \quad \text{and} \quad \left(\vec{\nabla}\tau\right)^2 = \frac{1}{\beta^2} , \tag{2.40}$$

introducing the P-wave velocity $\alpha = \sqrt{(\lambda + 2\mu)/\rho}$ and the S-wave velocity $\beta = \sqrt{\mu/\rho}$.

2.1.6 Solution of the eikonal equation

In Sections 2.1.3 and 2.1.5, I have shown that the eikonal equation has the same form for the acoustic as well as for the elastic case. The only difference is the inserted velocity, i. e., c, α, or β. Hence, they can be solved with a unified method. The eikonal equation can be rewritten in a more general form as

$$\left(\vec{\nabla}\tau\right)^2 = \frac{1}{v^2} \tag{2.41}$$

with $v = c$ for the acoustic case, $v = \alpha$ for P-wave velocity, and $v = \beta$ for S-wave velocity. Introducing the slowness vector $\vec{p}$ as the gradient of the traveltime τ into the characteristics equation (2.41) results in

$$\vec{p}^2 - \frac{1}{v^2} = 0 . \tag{2.42}$$

Please note the difference between:

$$\text{slowness vector} \quad : \quad \vec{p}(\vec{x}) = (p_1, p_2, p_3)^T = \vec{\nabla}\tau \quad \text{and}$$
$$\text{pressure} \quad : \quad p(\vec{x}) .$$

Equation (2.42) is a special case of the general form

$$\mathcal{H} = 0 , \tag{2.43}$$

where $\mathcal{H} = \mathcal{H}(\vec{x}, \vec{p}(\vec{x}))$ is the so-called Hamiltonian of the system. A curve $\vec{x} = \vec{x}(\xi)$ (ξ being a parameter increasing monotonously along the curve), along which equation (2.43) is satisfied, is called a characteristic of equation (2.43). However, in seismics based on ray theory, it is known as a ray. The parameter ξ depends on the particular form of the Hamiltonian $\mathcal{H}$ and cannot be arbitrarily chosen. It must satisfy

$$d\xi = \frac{d\tau}{\vec{p} \cdot \vec{\nabla}_{\vec{p}} \mathcal{H}} \qquad (2.44)$$

in all cases (see Bleistein, 1984). As mentioned above, there exist various different forms of the Hamiltonian $\mathcal{H}$, among which the following forms are the most commonly used ones:

$$\mathcal{H} = \frac{\vec{p} \cdot \vec{p} - \frac{1}{v^2}}{2}, \qquad (2.45a)$$

$$\mathcal{H} = (\vec{p} \cdot \vec{p})^{\frac{1}{2}} - \frac{1}{v}, \qquad (2.45b)$$

$$\mathcal{H} = \frac{1}{2} \ln \left(v^2 \vec{p} \cdot \vec{p} \right). \qquad (2.45c)$$

The corresponding parameter ξ is chosen as:

$$d\xi = v^2 d\tau = v\,ds = d\sigma, \qquad (2.46a)$$

$$d\xi = v d\tau = ds, \qquad (2.46b)$$

$$d\xi = d\tau, \qquad (2.46c)$$

respectively. Hereby, $ds = v d\tau$ has the obvious meaning of arclength. $d\sigma = v^2 d\tau$ is another parameter which is sometimes associated with the characteristical length scale along the ray. The most important application of this parameter can be found in the solution of transportation equations (see, e. g., Bleistein, 1986; Stockwell Jr., 1995).

Then, the characteristic, i. e., the ray, is with all this terminology the solution of the so-called characteristic system of first-order ordinary differential equations,

$$\frac{d\vec{x}}{d\xi} = \vec{\nabla}_{\vec{p}} \mathcal{H}, \qquad \frac{d\vec{p}}{d\xi} = -\vec{\nabla} \mathcal{H}, \qquad \text{and} \qquad \frac{d\tau}{d\xi} = \vec{p} \cdot \vec{\nabla}_{\vec{p}} \mathcal{H}. \qquad (2.47)$$

In other publications, the above equations can be found as the Hamiltonian canonical equations. Within the Hamilton formalism, $\vec{x}$ and $\vec{p}$ are treated as independent coordinates in a 6-dimensional phase space. Therefore, $\vec{x}$ and $\vec{p}$ are also called the canonical coordinates, and the vector $(\vec{x}, \vec{p})^T$ is called the canonical vector.

Equations (2.47) can be used to determine the ray trajectory. Once the ray trajectory is obtained, also the traveltime along the ray can be determined, provided the initial condition is given. Thus, it is referred to as the ray tracing system in ray theory.

2.1.7 Solution of the transport equation

In the following, I will refer to the transport equation of the acoustic case, i. e., equation (2.28), to find a solution. The elastic case is a bit more cumbersome and can be read in Červený (2001).

A family of rays, i. e., a ray field, is defined in such a way that each ray is orthogonal to the wavefront. This holds true only in the isotropic case, the anisotropic case will not be discussed here. If the ray field is regular, three parameters are introduced to uniquely identify each ray in such a ray field. The

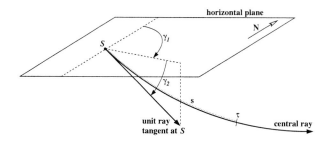

Figure 2.1: The ray coordinate system. For a wavefront emanating from a point source S in 3D, ray coordinates are given by the two take-off angles γ_1 and γ_2 and the eikonal τ. In this case, τ specifies a position of a point on the ray by its traveltime. Instead of τ, it is also possible to use the arclength s (dotted line).

condition of regularity depends on the local radius of curvature of the considered ray. As long as the distance to the described ray remains smaller than the local radius of curvature, a unique description is valid. Thus, the triplet $(\gamma_1, \gamma_2, \gamma_3)^T$ describes every point in the system reached by the wavefront, while it is most convenient to define γ_1 and γ_2 as the take-off angles of a ray emanating from a point source and γ_3 can be either the traveltime τ, i.e., the eikonal or the arclength s along the ray (see Figure 2.1). The relationship of the ray coordinates to the global Cartesian coordinates is given by

$$Q_{ij}^{(\vec{x})} = \frac{\partial x_i}{\partial \gamma_j}.$$
(2.48)

Now, the partial differential equation (2.28) can be rewritten as an ordinary differential equation along a ray using the ray coordinates. After some manipulations, the transport equation reads

$$\frac{2}{c^2}\frac{dA}{d\tau} + \frac{A}{c\mathcal{J}}\frac{d}{d\tau}\left(\frac{\mathcal{J}}{c}\right) = 0.$$
(2.49)

Here, $\mathcal{J}$ is called Jacobian determinant. As long as $\gamma_3 = s$, i.e., the arclength along the ray, $\mathcal{J}$ is also known as the ray Jacobian. $\mathcal{J}$ can be geometrically interpreted as the density of the ray field and can be expressed as a functional determinant (which is named after the mathematician Jacobi who first introduced it)

$$\mathcal{J} = \frac{1}{c}\left|\frac{d(x_1, x_2, x_3)}{d(\gamma_1, \gamma_2, \gamma_3)}\right|,$$
(2.50)

where x_1, x_2, x_3 are the global Cartesian coordinates and $\gamma_1, \gamma_2, \gamma_3$ or γ_1, γ_2, τ the ray coordinates. The Jacobian determinant provides important information about the behavior of vector-valued functions near a given point as it contains all first-order partial derivatives of the vector-valued function. Thus, it represents the best linear approximation to a differentiable function near a given point and can be used to express the change of amplitudes in the vicinity of a given point. However, the calculation of geometrical spreading out of $\mathcal{J}$ will not be further discussed here.

To solve equation (2.49), the variables can easily be separated. Finally, the main term for the amplitude is given by

$$A = \frac{\Sigma(\gamma_1, \gamma_2)}{\sqrt{\frac{\mathcal{J}}{c}}}$$
(2.51)

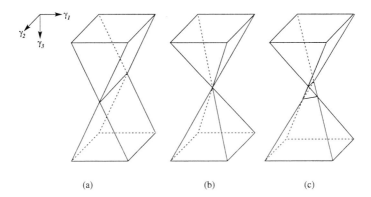

Figure 2.2: Some types of caustics. The radius of curvature becomes zero (a) for only one principal wavefront curvature, i. e., a first order caustic, (b) for both principal wavefront curvatures in the same point, i. e., a second order caustic, and (c) also for both principal wavefront curvatures but at different points, i. e., two consecutive first order caustics.

where $\sum$ stands for the constant of integration which depends only on the ray take-off parameters. The above solution of the transport equation implies, by now, the calculation of the Jacobian determinant and the search for initial ray data, i. e., the constant of integration. If the wavefront is known, the evaluation of $\mathcal{J}$ can be solved numerically (see Popov, 2002). How to obtain the initial ray data is discussed in Červený (2001). During the last years, methods were developed to directly estimate dynamic parameters from traveltime observations (see e. g. Gajewski, 1998, for details).

In the vicinity of so-called caustics, the Jacobian determinant $\mathcal{J}$ becomes zero because at least one of the principal wavefront curvature radii becomes zero, see Figure 2.2 or additionally Figure 2.4. Looking at the amplitude term (2.51), shows that the amplitude will tend to infinity the closer to the caustic point which is unphysical. Thus, for these points, the zero-order ray series ansatz breaks down as the term ΔA cannot be neglected anymore. Thus, some other expressions have to be found. The Gaussian beam method (see Popov, 2002) or the Maslov method are some commonly used methods to overcome the caustics problem.

2.2 The paraxial ray theory

In paraxial ray theory, two main terms have to be explained first namely paraxial and central ray. An arbitrarily chosen ray is hereby called the central ray and all other rays in its close vicinity are called paraxial rays. Then, the paraxial ray theory provides a linear description of the paraxial rays from the properties of a central ray. Thus, paraxial ray theory is used to derive quantities like the two-point traveltimes along the paraxial rays or the geometrical spreading factor along the central ray, etc. The derived traveltime formulae establish the fundamental basis of the common-reflection-surface (CRS) stack (see, e. g., Mann et al., 1999; Zhang et al., 2001).

The dynamic ray tracing system (2.47) can be expressed in many different forms and in various coordinate systems. The simplest expression is obtained in the ray-centered coordinate system q_1, q_2, and

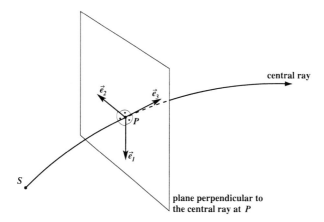

Figure 2.3: The ray-centered coordinate system. The vectors $\vec{e}_1$ and $\vec{e}_2$ are unit vectors located within the plane perpendicular to the ray in point P. The vector $\vec{e}_3$ is the unit ray tangent at P and coincides with the normal vector of the plane. The coordinates of point P are given by $(q_1 = 0, q_2 = 0, q_3)^T$. Hereby, point S denotes the origin of the ray-centered coordinate system with the coordinates $(0,0,0)^T$.

q_3 defined by the unit vectors $\vec{e}_1$, $\vec{e}_2$, and $\vec{e}_3$, respectively. Hereby, the ray-centered coordinate system is a curvilinear orthogonal coordinate system where the third axis, i. e., $\vec{e}_3$ is tangent to the ray itself. $\vec{e}_1$ and $\vec{e}_2$ are chosen in such a way that they are formed by two mutually perpendicular lines within a plane perpendicular to the ray at an arbitrary point P along the ray, see Figure 2.3. Thus, the unit vectors are constructed from the unit ray tangent, i. e., the unit normal of the perpendicular plane, the unit ray normal, and the unit ray binormal. The vectors $\vec{e}_1$ and $\vec{e}_2$ are not yet uniquely defined as they can be rotated within the plane perpendicular to the ray at the considered point P. How to choose the vectors $\vec{e}_1$ and $\vec{e}_2$ is described in more details in Popov and Pšenčík (1978). Furthermore, different notations in other coordinate system can be found in Červený (2001).

The eikonal equation expressed in ray-centered coordinates can be used to derive a simple system of ordinary linear differential equations of the first order for paraxial rays. The relevant system is called the paraxial ray tracing system. The ray tube is the volume around the central ray covered by the paraxial rays, see Figure 2.4. The dynamic ray tracing system can be immediately obtained from the paraxial ray tracing system. The dynamic ray tracing system consists of four ordinary linear differential equations of the first order. Both systems are closely connected and the difference will only show up in the different physical meaning of the calculated quantities (see, e. g., Červený, 2001).

The ray-centered coordinates q_1, q_2, and q_3 and the ray coordinates γ_1, γ_2, and τ can be incorporated in the relationship

$$\mathcal{J}|_{\gamma_1=\gamma_{10},\gamma_2=\gamma_{20}} = \left|\frac{d\,(x_1,x_2,x_3)}{d\,(\gamma_1,\gamma_2,\tau)}\right|_{\gamma_1=\gamma_{10},\gamma_2=\gamma_{20}} = v_0 \left|\frac{d\,(q_1,q_2)}{d\,(\gamma_1,\gamma_2)}\right|_{\gamma_1=\gamma_{10},\gamma_2=\gamma_{20}} \tag{2.52}$$

where the additional 0 in the index stands for parameters of the central ray. Inserting this into equa-

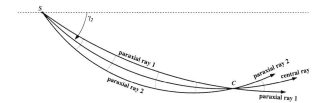

Figure 2.4: 2D sketch of a ray tube. γ_2 denotes the take-off angle of the central ray which is perturbed by $-/+d\gamma_2$ for paraxial ray 1 and 2, respectively. C depicts a caustic or focal point. In 3D, both principal wavefront curvature radii become zero in such a point.

tion (2.51) yields

$$A = \frac{\Sigma\left(\gamma_1, \gamma_2\right)}{\sqrt{\frac{1}{v}\left|\frac{d(q_1, q_2)}{d(\gamma_1, \gamma_2)}\right|}}\Bigg|_{\gamma_1=\gamma_{10}, \gamma_2=\gamma_{20}} . \tag{2.53}$$

With the additional notation

$$\begin{aligned}
Q_{1,1} &= \frac{\partial q_1}{\partial \gamma_1}\bigg|_{\gamma_1=\gamma_{10}, \gamma_2=\gamma_{20}} & Q_{1,2} &= \frac{\partial q_1}{\partial \gamma_2}\bigg|_{\gamma_1=\gamma_{10}, \gamma_2=\gamma_{20}} \\
Q_{2,1} &= \frac{\partial q_2}{\partial \gamma_1}\bigg|_{\gamma_1=\gamma_{10}, \gamma_2=\gamma_{20}} & Q_{2,2} &= \frac{\partial q_2}{\partial \gamma_2}\bigg|_{\gamma_1=\gamma_{10}, \gamma_2=\gamma_{20}}
\end{aligned} \tag{2.54}$$

introduced and defining the matrix

$$\mathbf{Q} = \left(\begin{array}{cc} Q_{1,1} & Q_{1,2} \\ Q_{2,1} & Q_{2,2} \end{array}\right), \tag{2.55}$$

the following expression for the amplitude (2.53) is obtained:

$$A = \frac{\Sigma_0\left(\gamma_1, \gamma_2\right)}{\sqrt{\frac{1}{v_0}\left|\det\left(\mathbf{Q}\right)\right|}} . \tag{2.56}$$

This expression for the amplitude is evaluated on the central ray of a ray tube (see, e. g., Popov, 2002). The conditions $\gamma_1 = \gamma_{10}$ and $\gamma_2 = \gamma_{20}$ could have also been considered as $q_1 = q_2 = 0$.

Now, the transformation matrix $\mathbf{P}$ from ray coordinates to covariant ray-centered components of the slowness vector $\vec{p}$ is introduced given by

$$\mathbf{P} = \left(\begin{array}{cc} P_{1,1} & P_{1,2} \\ P_{2,1} & P_{2,2}, \end{array}\right) \tag{2.57}$$

where the components are

$$\begin{aligned}
P_{1,1} &= \frac{\partial p_1}{\partial \gamma_1}\bigg|_{\gamma_1=\gamma_{10}, \gamma_2=\gamma_{20}} & P_{1,2} &= \frac{\partial p_1}{\partial \gamma_2}\bigg|_{\gamma_1=\gamma_{10}, \gamma_2=\gamma_{20}} \\
P_{2,1} &= \frac{\partial p_2}{\partial \gamma_1}\bigg|_{\gamma_1=\gamma_{10}, \gamma_2=\gamma_{20}} & P_{2,2} &= \frac{\partial p_2}{\partial \gamma_2}\bigg|_{\gamma_1=\gamma_{10}, \gamma_2=\gamma_{20}}
\end{aligned} . \tag{2.58}$$

With the matrix $\mathbf{V}$ composed of the second derivatives of the velocity v with respect to q along the central ray, i. e.,

$$\mathbf{V} = \left(v_{ij}\right) = \left(\frac{\partial^2 v}{\partial q_i \partial q_j}\bigg|_{q_1=q_2=0}\right) \qquad \text{for} \qquad i, j = 1, 2, \tag{2.59}$$

21

the dynamic ray tracing system can be written in the form

$$\frac{d\mathbf{Q}}{ds} = v_0 \mathbf{P}, \qquad \frac{d\mathbf{P}}{ds} = -\frac{1}{v_0^2} \mathbf{V}\mathbf{Q}, \tag{2.60}$$

where s is the arclength along the central ray (see, e. g., Červený, 2001). The reader should note that instead of the arclength s any other monotonic variable along the central ray, e. g., the traveltime, can be used.

2.2.1 Ray propagator matrix

As mentioned above, the properties of rays in the close vicinity of an arbitrarily chosen central ray can be described by the well-established paraxial ray theory (see, e. g., Červený, 2001; Bortfeld, 1989; Popov, 2002). With the paraxial approximation, the ray tracing system (2.47) is considered to be approximately valid in the close vicinity of the central ray (where it is exact). Furthermore, the ray propagator matrix plays an important role. A representation of the ray propagator matrix in the ray-centered coordinate system can be found in Červený (2001), Hubral (1983), and Popov (2002). Bortfeld (1989) has introduced a surface-to-surface propagator matrix which deals with a medium consisting of homogeneous, isotropic layers. A generalization to inhomogeneous, isotropic media can be found in Hubral et al. (1992a,b). The concept of anterior (usually the measurement surface, but not necessarily) and posterior surfaces (usually the measurement surface, but also the reflector itself or, in general, a totally different surface) was used in both cases to explain the physical meaning of the propagator matrix.

Equation (2.60) is a system of ordinary linear differential equations of the first order. To this system, there exist four linearly independent solutions. The fundamental matrix $\underline{\underline{\Pi}}$ of the system is called the propagator matrix of the dynamic ray tracing system or simply ray propagator matrix. The paraxial approximation implies that the dynamic parameters at an arbitrarily chosen point on a paraxial ray are linearly dependent on those at the corresponding initial point S on the central ray. One condition for the ray propagator matrix is to become the identity matrix at S on the central ray. Once the ray propagator matrix at S has been found along the ray, the solution of the dynamic ray tracing system for any initial condition at S is obtained merely by multiplying the ray propagator matrix by the matrix of the initial conditions (see Červený, 2001). Furthermore, kinematic parameters like the two-point traveltime of paraxial rays as well as dynamic parameters like the geometrical spreading factor can be obtained from the ray propagator matrix. Aspects concerning the amplitudes like the geometrical spreading factor will not be further investigated here, please refer to Červený (2001).

The 4×4 surface-to-surface ray propagator matrix $\underline{\underline{T}}$ which is employed in Bortfeld (1989) represents the propagator matrix in the global Cartesian coordinate system. The relationship between the $\underline{\underline{T}}$ and the $\underline{\underline{\Pi}}$ matrix has been published in Hubral et al. (1992a). Some useful properties of $\underline{\underline{T}}$ are explained in Appendix B. The 4×4 ray-centered propagator matrix $\underline{\underline{\Pi}}$ of a ray which connects an initial point S, i. e., the source point, and an end point R, i. e., the receiver point, can be written as

$$\underline{\underline{\Pi}}(S, R) = \begin{pmatrix} \mathbf{Q}_1 & \mathbf{Q}_2 \\ \mathbf{P}_1 & \mathbf{P}_2 \end{pmatrix}. \tag{2.61}$$

The 2×2 matrices $\mathbf{Q}_1, \mathbf{P}_1, \mathbf{Q}_2,$ and $\mathbf{P}_2$ can be obtained by solving the dynamic ray tracing system (2.60) for two initial conditions. The plane wave solution results from the initial conditions $\mathbf{Q}(S) = \mathcal{I}$ and

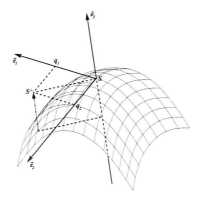

Figure 2.5: A wavefront which propagates in space in direction of the ray-centered unit vector $\vec{e}_3$. The wavefront crosses the point S on the central ray at the traveltime $\tau(S)$. The tangent plane spanned by the other two ray-centered unit vectors $\vec{e}_1$ and $\vec{e}_2$ contains the point $S^\perp$ on a paraxial ray. The distance in the $\vec{e}_3$-direction from the wavefront at $\tau(S)$ until the wavefront touches point $S^\perp$ at traveltime $\tau(S^\perp)$ is depicted by the dash-dotted arrow and given by $\frac{1}{2}\vec{q}^T\mathbf{K}(S)\vec{q}$.

$\mathbf{P}(S) = \mathbf{0}$ and the initial conditions $\mathbf{Q}(S) = \mathbf{0}$ and $\mathbf{P}(S) = \boldsymbol{I}$ lead to the point source solution (see Červený, 2001). The matrix $\boldsymbol{I}$ is the identity matrix (here: $i_{ij} = \delta_{ij}$ for $i, j = 1, 2$) and the matrix $\mathbf{0}$ is the zero matrix ($0_{ij} = 0$ for $i, j = 1, 2$).

For a source point S' and an end point R' on a paraxial ray in the vicinity of the central ray with points S and R, the deviations in location and slowness between R and R' are related to the deviations in location and slowness between S and S' given by

$$\begin{pmatrix} \Delta\vec{x}_R \\ \Delta\vec{p}_R \end{pmatrix} = \underline{\underline{\mathbf{\Pi}}}(S,R)\begin{pmatrix} \Delta\vec{x}_S \\ \Delta\vec{p}_S \end{pmatrix} \tag{2.62}$$

assuming a linear relationship, i. e., the paraxial approximation. Hereby, the ray propagator matrix $\underline{\underline{\mathbf{\Pi}}}$ also has the symplectic property described in Appendix B for the ray propagator matrix $\underline{\underline{\mathbf{T}}}$. Briefly formulated, it is expressed by

$$\det\left(\underline{\underline{\mathbf{\Pi}}}\right) = 1. \tag{2.63}$$

2.2.2 Paraxial traveltimes

One of the most important applications of the ray propagator matrix lies in the derivation of the traveltimes of the paraxial rays. The first step is to utilize the matrix of second derivatives of traveltime in ray-centered coordinates q_i for $i = 1, 2$. Here, a point on the central ray is denoted S with $q_3 = s$. Then, the components of the matrix of the second derivatives in ray-centered coordinates are given by

$$\left(\frac{\partial^2\tau}{\partial q_i \partial q_j}\right) = \left(\frac{\partial}{\partial q_j}\left(\frac{\partial\tau}{\partial q_i}\right)\right) = \left(\frac{\partial p_i}{\partial q_j}\right) = \left(\frac{\partial p_i}{\partial\gamma_k}\frac{\partial\gamma_k}{\partial q_j}\right) = \mathbf{PQ}^{-1} \equiv \mathbf{M}(S) \qquad \text{with} \qquad i, j, k = 1, 2. \tag{2.64}$$

I defined $\mathbf{M}(S)$ to express the traveltime from a starting point S on the central ray to a point $S^{\perp}$ on a paraxial ray, which is assumed to be located upon a plane tangent to the wavefront at S. Then, the traveltime reads

$$\tau\left(S^{\perp}\right) = \tau(S) + \frac{1}{2}\vec{q}^{\,T}\mathbf{M}(S)\vec{q}, \tag{2.65}$$

where $\vec{q} = \vec{q}(S^{\perp})$ with $\vec{q} = (q_1, q_2)^T$. Thus, the slowness vector $\vec{p} = \vec{p}(S^{\perp}) = (p_1, p_2)^T$ at point $S^{\perp}$ can be approximated by

$$p_i = \frac{\partial \tau(S^{\perp})}{\partial q_i} = m_{ij}(S)q_j \qquad \text{with} \qquad i, j = 1, 2. \tag{2.66}$$

To explain the meaning of equation (2.65), I refer to Figure 2.5. There, a wavefront is supposed to propagate along the unit vector $\vec{e}_3$ in ray-centered coordinates. The wavefront arrives at point S at traveltime $\tau(S)$. The plane spanned by the other two ray-centered unit vectors $\vec{e}_1$ and $\vec{e}_2$, i.e., the tangent plane at point S, contains another point $S^{\perp}$ with the coordinates $\vec{q} = (q_1, q_2)^T$. The distance from point $S^{\perp}$ to the wavefront at $\tau(S)$ in the $\vec{e}_3$-direction can be calculated as $\left(\vec{q}^{\,T}\mathbf{K}(S)\vec{q}\right)/2$. The matrix $\mathbf{K}(S)$ is the curvature matrix which is given by

$$\mathbf{K}(S) = \mathbf{M}(S)v_S \tag{2.67}$$

where v_S is the velocity at point S. Dividing this distance by the velocity v_S yields the time delay that the wavefront needs to reach point $S^{\perp}$, namely $\left(\vec{q}^{\,T}\mathbf{M}(S)\vec{q}\right)/2$. Thus, the traveltime in point $S^{\perp}$ can be written as in equation (2.65).

For the further derivations, it is important to discuss the dynamic evolution of the matrix $\mathbf{M}(S) = \mathbf{M}$ along the central ray. The differentiation with respect to τ reads

$$\frac{d\mathbf{M}}{d\tau} = \frac{d\mathbf{P}}{d\tau}\mathbf{Q}^{-1} + \mathbf{P}\frac{d\mathbf{Q}^{-1}}{d\tau}. \tag{2.68}$$

To simplify the second term of this equation, the following relations can be taken into account:

$$\mathbf{Q}\mathbf{Q}^{-1} = \mathbf{I}. \tag{2.69}$$

Differentiating both sides with respect to τ yields

$$\frac{d\mathbf{Q}}{d\tau}\mathbf{Q}^{-1} + \mathbf{Q}\frac{d\mathbf{Q}^{-1}}{d\tau} = \mathbf{0} \tag{2.70}$$

which implies

$$\frac{d\mathbf{Q}^{-1}}{d\tau} = -\mathbf{Q}^{-1}\frac{d\mathbf{Q}}{d\tau}\mathbf{Q}^{-1}. \tag{2.71}$$

Furthermore, as the first two components of the slowness vector $\vec{p}$ vanish, the following can be stated:

$$\frac{\partial}{\partial q_i}\left(\frac{1}{v}\right) = -\frac{1}{v^2}\frac{\partial v}{\partial q_i} = -\frac{1}{v^2}\frac{\partial}{\partial q_j}\frac{\partial v}{\partial q_i}\bigg|_{\vec{q}=0}q_j \equiv -\frac{1}{v^2}v_{ij}q_j \qquad \text{with} \qquad i, j = 1, 2, \tag{2.72}$$

where v_{ij} denotes the elements of the matrix $\mathbf{V}$ given by definition (2.59) which I just repeat here for convenience:

$$\mathbf{V} = \left(v_{ij}\right) = \left(\frac{\partial^2 v}{\partial q_i \partial q_j}\bigg|_{\vec{q}=0}\right) \qquad \text{with} \qquad i, j = 1, 2.$$

With relations (2.71) and (2.72), equation (2.68) can be rewritten in the form

$$\frac{d\mathbf{M}}{d\tau} + \frac{1}{v}\mathbf{V} + v^2\mathbf{M}^2 = 0 \tag{2.73}$$

which is a non-linear partial differential equation of first order.

The dynamic ray tracing system (2.60) can also be written as

$$\frac{d}{ds}\begin{pmatrix} \mathbf{Q} \\ \mathbf{P} \end{pmatrix} = \underline{\underline{\mathbf{W}}}\begin{pmatrix} \mathbf{Q} \\ \mathbf{P} \end{pmatrix}, \tag{2.74}$$

and with the relation $ds = vd\tau$, the dynamic ray tracing system reads

$$\frac{d}{d\tau}\begin{pmatrix} \mathbf{Q} \\ \mathbf{P} \end{pmatrix} = v\underline{\underline{\mathbf{W}}}\begin{pmatrix} \mathbf{Q} \\ \mathbf{P} \end{pmatrix} \tag{2.75}$$

where

$$\underline{\underline{\mathbf{W}}} = \begin{pmatrix} \mathbf{0} & v\mathbf{I} \\ -\frac{1}{v^2}\mathbf{V} & \mathbf{0} \end{pmatrix} = \begin{pmatrix} 0 & 0 & v & 0 \\ 0 & 0 & 0 & v \\ -\frac{1}{v^2}\frac{\partial^2 v}{\partial q_1^2}\big|_{\vec{q}=0} & -\frac{1}{v^2}\frac{\partial^2 v}{\partial q_1 \partial q_2}\big|_{\vec{q}=0} & 0 & 0 \\ -\frac{1}{v^2}\frac{\partial^2 v}{\partial q_2 \partial q_1}\big|_{\vec{q}=0} & -\frac{1}{v^2}\frac{\partial^2 v}{\partial q_2^2}\big|_{\vec{q}=0} & 0 & 0 \end{pmatrix}. \tag{2.76}$$

With equations (2.73), (2.74), and (2.75), the matrix $\mathbf{M}(S)$ can be determined at any arbitrary point S along the central ray if either the matrices $\mathbf{Q}$ and $\mathbf{P}$ or the matrix $\mathbf{M}$ are known at any arbitrary initial point S_0 along the central ray.

To express now the traveltime in ray-centered coordinates, I assume the matrix $\mathbf{M}$ to be known at any point along the central ray. Consider a point S' with the ray-centered coordinates $(q_1, q_2, s + \Delta s)^T$ in the close vicinity of the central ray, i. e., $q_1 \neq 0$ and $q_2 \neq 0$. Then, the traveltime along a paraxial ray passing through point S' can be approximated from the traveltime at a point S with the ray-centered coordinates $(0, 0, s)^T$ as

$$\tau(S') = \tau(S) + \frac{\partial \tau}{\partial q_i}\bigg|_S q_i + \frac{\partial \tau}{\partial s}\bigg|_S \Delta s + \frac{1}{2}\frac{\partial^2 \tau}{\partial q_i \partial q_j}\bigg|_S q_i q_j + \frac{1}{2}\frac{\partial^2 \tau}{\partial s^2}\bigg|_S \Delta s^2 \qquad i, j = 1, 2. \tag{2.77}$$

The projection of the slowness vector onto the plane tangent to the wavefront vanishes according to the definition of the ray-centered coordinate system. This can also be stated when taking into account that the dynamic ray tracing system for the third component of ray-centered coordinates implies $p_3 = 1/v$, which yields

$$\frac{\partial \tau}{\partial q_i}\bigg|_S = 0 \qquad \text{for} \qquad i = 1, 2. \tag{2.78}$$

$p_3 = 1/v$ also implies that

$$\frac{\partial \tau}{\partial s}\bigg|_S = \frac{1}{v(S)} \qquad \text{and} \qquad \frac{\partial^2 \tau}{\partial s^2}\bigg|_S = -\frac{1}{v^2(S)}\frac{\partial v(S)}{\partial s}. \tag{2.79}$$

Thus, inserting the equivalence from equation (2.64) and equations (2.78) and (2.79) into equation (2.77), the approximation of the paraxial traveltime reads

$$\tau(S') = \tau(S) + \frac{1}{v(S)}\Delta s - \frac{1}{2v^2(S)}\frac{\partial v(S)}{\partial s}\Delta s^2 + \frac{1}{2}m_{ij}(S)q_i q_j \qquad \text{for} \qquad i, j = 1, 2. \tag{2.80}$$

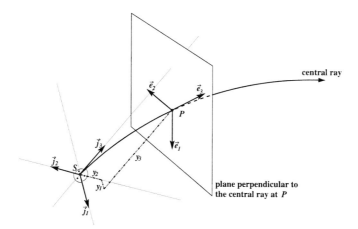

Figure 2.6: The local ray-centered Cartesian coordinate system. The unit vectors $\vec{j}_1$, $\vec{j}_2$, and $\vec{j}_3$ are identical with the unit vectors $\vec{e}_1$, $\vec{e}_2$, and $\vec{e}_3$ of the ray-centered coordinate system but only in the origin S with the coordinates $(0,0,0)^T$ in both systems. Point P has the local ray-centered Cartesian coordinates $(y_1,y_2,y_3)^T$. The quantities y_1, y_2, and y_3 are depicted as dash-dotted, dash-double-dotted, and dash-triple-dotted lines, respectively. While the ray-centered unit vectors change with the position along the ray, the local ray-centered Cartesian unit vectors remain the same as they are at the origin S.

2.2.3 Paraxial traveltimes in local ray-centered Cartesian coordinates

Many practical applications demand in- and output in Cartesian coordinates. In order to provide a traveltime formula convenient for a more practical purpose, I introduce the local ray-centered Cartesian coordinate system. The local ray-centered Cartesian coordinate system has its origin S on the central ray. The unit vectors $\vec{e}_1$, $\vec{e}_2$, and $\vec{e}_3$ of the ray-centered coordinate system are identical with the unit vectors $\vec{j}_1$, $\vec{j}_2$, and $\vec{j}_3$ of the local ray-centered Cartesian coordinate system only in the origin S. The coordinates of an arbitrary point S' or P are defined as

$$\vec{y} = \begin{pmatrix} y_1 \\ y_2 \\ y_3 \end{pmatrix}. \tag{2.81}$$

The unit vector $\vec{e}_3$ of the ray-centered coordinate system is always tangent to the central ray. In contrast to that, the unit vector $\vec{j}_3$ is only tangent to the central ray in its origin S. Thus, $q_3 = s$ is the arclength along the central ray, whereas y_3 is measured along the tangent to the central ray at the origin S. The coordinates of an arbitrary point S' can be written as

$$\vec{r}(\vec{y})\big|_{S'} = \vec{r}\big|_S + y_1\vec{j}_1 + y_2\vec{j}_2 + y_3\vec{j}_3 \tag{2.82}$$

where $\vec{r}\big|_S$ denotes the position of the origin S. The perturbation of the vector reads by definition

$$d\vec{r} = dy_1\vec{j}_1 + dy_2\vec{j}_2 + dy_3\vec{j}_3. \tag{2.83}$$

The coordinates of point S' in ray-centered coordinates are given by

$$\vec{r}(q_1, q_2, s)\big|_{S'} = q_1\vec{e}_1 + q_2\vec{e}_2 + \vec{r}(0, 0, s)\big|_P \qquad (2.84)$$

where $\vec{r}(0, 0, s)\big|_P$ is the location of the origin in the ray-centered coordinate system. Please note that point S and P are the same when $q_3 = s \equiv 0$. The perturbation of this vector in the ray-centered coordinate system then reads

$$
\begin{aligned}
d\vec{r} &= dq_1\vec{e}_1 + dq_2\vec{e}_2 + \left(\frac{d}{ds}\vec{r}(0,0,s)\big|_P + q_1\frac{d}{ds}\vec{e}_1 + q_2\frac{d}{ds}\vec{e}_2\right)ds \\
&= dq_1\vec{e}_1 + dq_2\vec{e}_2 + \left(1 + \left(\frac{1}{v}\frac{\partial v}{\partial q_i}\right)\big|_{q_1=0,q_2=0} q_i\right)ds\vec{e}_3 \qquad \text{with} \qquad i = 1, 2.
\end{aligned}
\qquad (2.85)
$$

Comparing the coefficients of equations (2.83) and (2.85) yields

$$dy_1 = dq_1, \qquad (2.86a)$$

$$dy_2 = dq_2, \qquad (2.86b)$$

$$dy_3 = \left(1 + \left(\frac{1}{v}\frac{\partial v}{\partial q_i}\right)\Big|_{q_1=0,q_2=0} q_i\right)ds \equiv h\,ds \qquad (2.86c)$$

with

$$h = 1 + \left(\frac{1}{v}\frac{\partial v}{\partial q_i}\right)\Big|_{q_1=0,q_2=0} q_i \qquad \text{with} \qquad i = 1, 2. \qquad (2.87)$$

To express the paraxial traveltime formula (2.80) in the local ray-centered Cartesian coordinate system, the relations of two points $S = (0,0,0)^T$ and $S' = (y_1, y_2, y_3)^T$ given by equations (2.86) can be incorporated in the following form:

$$\frac{\partial}{\partial s} = h\frac{\partial}{\partial y_3}, \qquad \Delta s = h^{-1}y_3 \approx \left(1 - \frac{1}{v}\frac{\partial v}{\partial y_i}\right)y_3, \qquad \Delta s^2 \approx y_3^2, \qquad \text{and} \qquad m_{ij}^{(y)} = m_{ij} \qquad (2.88)$$

where $m_{ij}^{(y)} = \underline{\mathbf{M}}^{(y)}(S)$, $(i, j = 1, 2)$ are the upper left 2×2 submatrix elements of the 3×3 matrix of second derivatives in the local ray-centered Cartesian coordinate system. Substituting these relations into equation (2.80) yields

$$\tau(S') = \tau(S) + \frac{1}{v(S)}h^{-1}y_3 - \frac{1}{2v^2(S)}h\frac{\partial v(S)}{\partial y_3}y_3^2 + \frac{1}{2}m_{ij}(S)y_i y_j \qquad i, j = 1, 2. \qquad (2.89)$$

Resubstituting h from equation (2.87) and omitting third-order terms yields

$$\tau(S') = \tau(S) + \frac{1}{v(S)}y_3 - \frac{1}{2v^2(S)}\frac{\partial v(S)}{\partial y_3}y_3^2 - \frac{1}{v^2(S)}\frac{\partial v(S)}{\partial y_i}y_i y_3 + \frac{1}{2}m_{ij}(S)y_i y_j \qquad i, j = 1, 2, \qquad (2.90)$$

or, in a more compact form,

$$\tau(S') = \tau(S) + \vec{y}^T\vec{p}^{(y)}(S) + \frac{1}{2}\vec{y}^T\underline{\mathbf{M}}^{(y)}(S)\vec{y}, \qquad (2.91)$$

i. e., the paraxial traveltime formula in the local ray-centered coordinate system. $\vec{y}$ is given by equation (2.81),

$$\vec{p}^{(y)}(S) = \begin{pmatrix} p_1^{(y)}(S) \\ p_2^{(y)}(S) \\ p_3^{(y)}(S) \end{pmatrix} = \begin{pmatrix} 0 \\ 0 \\ \frac{1}{v(S)} \end{pmatrix}, \qquad (2.92)$$

and

$$\underline{\mathbf{M}}^{(y)}(S) = \begin{pmatrix} m_{11}^{(y)} & m_{12}^{(y)} & -\frac{1}{v^2}\frac{\partial v}{\partial y_1} \\ m_{21}^{(y)} & m_{22}^{(y)} & -\frac{1}{v^2}\frac{\partial v}{\partial y_2} \\ -\frac{1}{v^2}\frac{\partial v}{\partial y_1} & -\frac{1}{v^2}\frac{\partial v}{\partial y_2} & -\frac{1}{v^2}\frac{\partial v}{\partial y_3} \end{pmatrix}_S . \qquad (2.93)$$

27

2.2.4 The two-point eikonal

Consider two points S and R on a central ray and in its vicinity points S' and R' on a paraxial ray. Furthermore, the traveltime and the ray propagator matrix are assumed to be known along the central ray from S to R in ray-centered coordinates. Then, the approximated traveltime along the paraxial ray passing through S' and R' can be computed from the central ray and is called the two-point eikonal. To calculate the two-point eikonal, two more points have to be considered: $S^\perp$ and $R^\perp$ are the intersection points of the paraxial ray with the tangent plane to the wavefront passing through S and R, respectively. The coordinates of $S^\perp$ and $R^\perp$ in the ray-centered coordinate system are denoted by

$$\vec{q}\left(S^\perp\right) = \begin{pmatrix} q_1\left(S^\perp\right) \\ q_2\left(S^\perp\right) \end{pmatrix} \qquad \text{and} \qquad \vec{q}\left(R^\perp\right) = \begin{pmatrix} q_1\left(R^\perp\right) \\ q_2\left(R^\perp\right) \end{pmatrix}. \tag{2.94}$$

Thus, the points S' and R' have the coordinates in the local ray-centered Cartesian coordinate system

$$\vec{y}(S') = \begin{pmatrix} y_1\,(S') = q_1\left(S^\perp\right) \\ y_2\,(S') = q_2\left(S^\perp\right) \\ y_3\,(S') \end{pmatrix} \qquad \text{and} \qquad \vec{y}(R') = \begin{pmatrix} y_1\,(R') = q_1\left(R^\perp\right) \\ y_2\,(R') = q_2\left(R^\perp\right) \\ y_3\,(R') \end{pmatrix}. \tag{2.95}$$

Formulating the traveltime difference using equation (2.66) yields

$$\begin{aligned}
\Delta\tau &= \Delta\tau\,(R, R') - \Delta\tau\,(S, S') \\
&= (\tau\,(R') - \tau\,(R)) - (\tau\,(S') - \tau\,(S)) \\
&= \left(\frac{y_3\,(R')}{v\,(R)} - \frac{1}{2v^2\,(R)}\frac{\partial v\,(R)}{\partial y_3}y_3\,(R')^2 - \frac{1}{v^2\,(R)}\frac{\partial v\,(R)}{\partial y_i}y_i\,(R')\,y_3\,(R') + \frac{1}{2}y_i\,(R')\,p_i\left(R^\perp\right) \right) \\
&\quad - \left(\frac{y_3\,(S')}{v\,(S)} - \frac{1}{2v^2\,(S)}\frac{\partial v\,(S)}{\partial y_3}y_3\,(S')^2 - \frac{1}{v^2\,(S)}\frac{\partial v\,(S)}{\partial y_i}y_i\,(S')\,y_3\,(S') + \frac{1}{2}y_i\,(S')\,p_i\left(S^\perp\right) \right) \\
&\hspace{8cm} \text{with} \quad i = 1, 2. \tag{2.96}
\end{aligned}$$

The linear relationship set up by the ray propagator matrix in the ray-centered coordinate system

$$\begin{pmatrix} \vec{q}\left(R^\perp\right) \\ \vec{p}\left(R^\perp\right) \end{pmatrix} = \underline{\underline{\mathbf{\Pi}}}\,(S, R)\begin{pmatrix} \vec{q}\left(S^\perp\right) \\ \vec{p}\left(S^\perp\right) \end{pmatrix} \tag{2.97}$$

can be rewritten as

$$\vec{q}\left(R^\perp\right) = \mathbf{Q}_1\vec{q}\left(S^\perp\right) + \mathbf{Q}_2\vec{p}\left(S^\perp\right) \qquad \text{and} \tag{2.98a}$$

$$\vec{p}\left(R^\perp\right) = \mathbf{P}_1\vec{q}\left(S^\perp\right) + \mathbf{P}_2\vec{p}\left(S^\perp\right) \tag{2.98b}$$

,in which

$$\vec{p}\left(S^\perp\right) = \begin{pmatrix} p_1\left(S^\perp\right) \\ p_2\left(S^\perp\right) \end{pmatrix} \qquad \text{and} \qquad \vec{p}\left(R^\perp\right) = \begin{pmatrix} p_1\left(R^\perp\right) \\ p_2\left(R^\perp\right) \end{pmatrix}. \tag{2.99}$$

Thus, if there is a point source at S, the first terms of equations (2.98) vanish. If $\vec{p}\left(S^\perp\right)$ are taken as the initial ray coordinates, the matrix of second derivatives of traveltimes at R due to the point source at S can easily be computed as

$$m_{ij}\,(S, R) = \frac{\partial p_i\left(R^\perp\right)}{\partial p_k\left(S^\perp\right)}\frac{\partial p_k\left(S^\perp\right)}{\partial q_j\left(R^\perp\right)} = \left(\mathbf{P}_2\mathbf{Q}_2^{-1}\right)_{ij} \qquad \text{with} \qquad i, j, k = 1, 2, \tag{2.100}$$

which implies

$$\mathbf{M}(S,R) = \mathbf{P}_2 \mathbf{Q}_2^{-1}. \tag{2.101}$$

Now, the matrix of the second derivatives of traveltime at S due to a point source at R can be similarly calculated as

$$\mathbf{M}(R,S) = \mathbf{Q}_1^T \mathbf{Q}_2^{-1T} = \mathbf{Q}_2^{-1} \mathbf{Q}_1. \tag{2.102}$$

Those derivatives are useful for solving the two equations from system (2.98) for $\vec{p}(S^\perp)$ and $\vec{p}(R^\perp)$. Starting with the solution for $\vec{p}(S^\perp)$ and inserting this solution into the solution for $\vec{p}(R^\perp)$ yields

$$\vec{p}(S^\perp) = -\mathbf{M}(R,S)\vec{q}(S^\perp) + \mathbf{Q}_2\vec{q}(R^\perp) \quad \text{and} \tag{2.103}$$

$$\vec{p}(R^\perp) = -\mathbf{Q}_2^{-1T}\vec{q}(S^\perp) + \mathbf{M}(S,R)\vec{q}(R^\perp). \tag{2.104}$$

Inserting the above expressions for the slowness at $S^\perp$ and $R^\perp$ in ray-centered coordinates into the traveltime difference (2.96) leads to

$$
\begin{aligned}
\Delta\tau = & \left(\frac{y_3(R')}{v(R)} - \frac{1}{2v^2(R)}\frac{\partial v(R)}{\partial y_3} y_3(R')^2 - \frac{1}{v^2(R)}\frac{\partial v(R)}{\partial y_i} y_i(R') y_3(R') \right) \\
& - \left(\frac{y_3(S')}{v(S)} - \frac{1}{2v^2(S)}\frac{\partial v(S)}{\partial y_3} y_3(S')^2 - \frac{1}{v^2(S)}\frac{\partial v(S)}{\partial y_i} y_i(S') y_3(S') \right) \\
& + \frac{1}{2} y_i(R') m_{ij}(S,R) y_j(R') + \frac{1}{2} y_i(S') m_{ij}(R,S) y_j(S') \\
& - \frac{1}{2} y_i(R') \left(\mathbf{Q}_2^{-1T} \right)_{ij} y_j(S') - \frac{1}{2} y_i(S') \left(\mathbf{Q}_2^{-1} \right)_{ij} y_j(R')
\end{aligned}
\qquad i = 1,2, \tag{2.105}
$$

or, expressed in a shorter form using vectors and matrices and combining the last two terms as they are identical,

$$
\begin{aligned}
\Delta\tau = & \ \vec{y}^T(R')\vec{p}^{(y)}(R) \\
& -\vec{y}^T(S')\vec{p}^{(y)}(S) \\
& +\frac{1}{2}\vec{y}^T(R')\underline{\mathbf{M}}(S,R)\vec{y}(R') + \frac{1}{2}\vec{y}^T(S')\underline{\mathbf{M}}(R,S)\vec{y}(S') \\
& -\vec{y}^T(S')\underline{\mathbf{R}}(S,R)\vec{y}(R'),
\end{aligned}
\tag{2.106}
$$

with

$$\underline{\mathbf{M}}(S,R) = \begin{pmatrix} m_{11}(S,R) & m_{12}(S,R) & -\frac{1}{v^2(R)}\frac{\partial v(R)}{\partial y_1} \\ m_{21}(S,R) & m_{22}(S,R) & -\frac{1}{v^2(R)}\frac{\partial v(R)}{\partial y_2} \\ -\frac{1}{v^2(R)}\frac{\partial v(R)}{\partial y_1} & -\frac{1}{v^2(R)}\frac{\partial v(R)}{\partial y_2} & -\frac{1}{v^2(R)}\frac{\partial v(R)}{\partial y_3} \end{pmatrix}, \tag{2.107a}$$

$$\underline{\mathbf{M}}(R,S) = \begin{pmatrix} m_{11}(R,S) & m_{12}(R,S) & -\frac{1}{v^2(S)}\frac{\partial v(S)}{\partial y_1} \\ m_{21}(R,S) & m_{22}(R,S) & -\frac{1}{v^2(S)}\frac{\partial v(S)}{\partial y_2} \\ -\frac{1}{v^2(S)}\frac{\partial v(S)}{\partial y_1} & -\frac{1}{v^2(S)}\frac{\partial v(S)}{\partial y_2} & -\frac{1}{v^2(S)}\frac{\partial v(S)}{\partial y_3} \end{pmatrix}, \quad \text{and} \tag{2.107b}$$

$$\underline{\mathbf{R}}(S,R) = \begin{pmatrix} \mathbf{Q}_2^{-1} & & 0 \\ & & 0 \\ 0 & 0 & 0 \end{pmatrix}. \tag{2.107c}$$

To finally transform the traveltime difference formula into the newly deduced local ray-centered Cartesian coordinate system, two local Cartesian coordinate systems are established, one at S and the other at R:

$$\vec{x}(S) = \begin{pmatrix} x_1(S) \\ x_2(S) \\ x_3(S) \end{pmatrix} \quad \text{and} \quad \vec{x}(R) = \begin{pmatrix} x_1(R) \\ x_2(R) \\ x_3(R) \end{pmatrix}, \tag{2.108}$$

while the third component of both systems is pointing into depth. The origin of the local Cartesian coordinate system at S is set to be the source point and the azimuth angles of the incidence ray at S,

29

i. e., on the central ray, are denoted with (β_S, θ_S). The displacement of a point S' in the close vicinity of the source point S with respect to S is given by $(x_1(S'), x_2(S'), x_3(S'))^T$. The origin of the local Cartesian coordinate system at R coincides with the receiver point along the central ray. Here, the azimuth angles are denoted with (β_R, θ_R) and the displacement of a point R' with respect to R is given by $(x_1(R'), x_2(R'), x_3(R'))^T$. Thus, the coordinate transformations at the source point S and the receiver point R can be written as

$$
\begin{pmatrix} y_1(S') \\ y_2(S') \\ y_3(S') \end{pmatrix} = \underline{\mathbf{D}}(S) \begin{pmatrix} x_1(S') \\ x_2(S') \\ x_3(S') \end{pmatrix} \quad \text{and} \quad \begin{pmatrix} y_1(R') \\ y_2(R') \\ y_3(R') \end{pmatrix} = \underline{\mathbf{D}}(R) \begin{pmatrix} x_1(R') \\ x_2(R') \\ x_3(R') \end{pmatrix} \tag{2.109}
$$

where the combined rotation matrices with respect to β and θ depending on the source or receiver point $\underline{\mathbf{D}}(S)$ and $\underline{\mathbf{D}}(R)$ are given by

$$
\underline{\mathbf{D}}(S) = \begin{pmatrix} \cos\theta_S \cos\beta_S & \sin\theta_S \cos\beta_S & -\sin\beta_S \\ -\sin\theta_S & \cos\theta_S & 0 \\ \cos\theta_S \sin\beta_S & \sin\theta_S \sin\beta_S & \cos\beta_S \end{pmatrix} \quad \text{and} \tag{2.110a}
$$

$$
\underline{\mathbf{D}}(R) = \begin{pmatrix} \cos\theta_R \cos\beta_R & \sin\theta_R \cos\beta_R & -\sin\beta_R \\ -\sin\theta_R & \cos\theta_R & 0 \\ \cos\theta_R \sin\beta_R & \sin\theta_R \sin\beta_R & \cos\beta_R \end{pmatrix} . \tag{2.110b}
$$

To obtain the final expression of the two-point eikonal in the local Cartesian coordinate system, equations (2.109) have to be substituted into equation (2.106). This is the expression of the moveout formula for the paraxial rays in the local Cartesian coordinate system and can be used to formulate the traveltime formula for paraxial rays:

$$
\begin{aligned}
\tau(S', R') = & \ \tau(S, R) + \Delta\tau \\
= & \ \tau(S, R) + \vec{x}^T(R') \underline{\mathbf{D}}^T(R) \vec{p}^{(y)}(R) - \vec{x}^T(S') \underline{\mathbf{D}}^T(S) \vec{p}^{(y)}(S) \\
& + \tfrac{1}{2} \vec{x}^T(R') \underline{\mathbf{D}}^T(R) \underline{\mathbf{M}}(S, R) \underline{\mathbf{D}}(R) \vec{x}(R') \\
& + \tfrac{1}{2} \vec{x}^T(S') \underline{\mathbf{D}}^T(S) \underline{\mathbf{M}}(R, S) \underline{\mathbf{D}}(S) \vec{x}(S') \\
& - \vec{x}^T(S') \underline{\mathbf{D}}^T(S) \underline{\mathbf{R}}(S, R) \underline{\mathbf{D}}(R) \vec{x}(R') .
\end{aligned} \tag{2.111}
$$

For a planar measurement surface, Ursin (1982) suggested that in practice a hyperbolic form of the traveltime approximation is a better approximation to the real traveltime response than the parabolic approximation given by equation (2.111). This was later also verified by the work of Höcht (1998), Jäger (1999), Müller (1999), and Bergler (2001). The hyperbolic traveltime formula is calculated from the parabolic form by taking the square of both sides of equation (2.111) and retaining only terms up to second order. Then, the hyperbolic form of the traveltime approximation reads

$$
\begin{aligned}
\tau(S', R')^2 = & \ \left(\tau(S, R) + \vec{x}^T(R') \vec{p}^{(x)}(R) - \vec{x}^T(S') \vec{p}^{(x)}(S) \right)^2 \\
& + \tau(S, R) \vec{x}^T(R') \underline{\mathbf{M}}^{(x)}(S, R) \vec{x}(R') \\
& + \tau(S, R) \vec{x}^T(S') \underline{\mathbf{M}}^{(x)}(R, S) \vec{x}(S') \\
& - 2\tau(S, R) \vec{x}^T(S') \underline{\mathbf{R}}^{(x)}(S, R) \vec{x}(R')
\end{aligned} \tag{2.112}
$$

where the slowness vectors $\vec{p}^{(x)}(S)$ and $\vec{p}^{(x)}(R)$ are defined as

$$
\vec{p}^{(x)}(S) = \underline{\mathbf{D}}^T(S) \vec{p}^{(y)}(S) \quad \text{and} \quad \vec{p}^{(x)}(R) = \underline{\mathbf{D}}^T(R) \vec{p}^{(y)}(R) \tag{2.113}
$$

and the expressions of the matrices of second derivatives of traveltime are shortened by combining them with the corresponding rotation matrices:

$$\underline{\mathbf{M}}^{(x)}(S,R) = \underline{\mathbf{D}}^T(R)\underline{\mathbf{M}}(S,R)\underline{\mathbf{D}}(R), \tag{2.114}$$

$$\underline{\mathbf{M}}^{(x)}(R,S) = \underline{\mathbf{D}}^T(S)\underline{\mathbf{M}}(R,S)\underline{\mathbf{D}}(S), \tag{2.115}$$

$$\underline{\mathbf{R}}^{(x)}(S,R) = \underline{\mathbf{D}}^T(S)\underline{\mathbf{R}}(S,R)\underline{\mathbf{D}}(R). \tag{2.116}$$

Chapter 3

Common-Reflection-Surface stack

In this chapter, I will present two different ways to derive the traveltime formulae on which the Common-Reflection-Surface (CRS) stack is based. Firstly, I start with a geometrical derivation of the traveltime formula for simulating a zero-offset (ZO) section from 2D plane measurement surface recordings. Secondly, I will make use of the traveltime formula derived in Chapter 2 to obtain a more complex description of traveltime surfaces which is suited for simulating again a ZO section but now for measurements along a rough top-surface topography.

3.1 Wave propagation in 2.5D media

The formulae derived in the previous chapter are all valid for 3D wave propagation in an arbitrary 3D medium. Such a 3D medium is usually approximated by a 3D model which varies arbitrarily in all three coordinate directions. In many cases, a so-called 2.5D medium can also be used to describe the subsurface structure. For such a 2.5D model the 3D medium has to be symmetrical to a plane which contains the unit vector in x_3-direction (the x_3-direction is usually associated with the depth axis). Such a plane of symmetry can contain an arbitrary vector in the x_1-x_2 plane. To simplify matters, an appropriate transformation rotates the coordinates in such a way that the plane of symmetry is the x_1-x_3 plane or any other plane parallel to the x_1-x_3 plane. Thus, the elastic properties in the x_2-direction are constant.

To be at least able to fully image an arbitrary 3D subsurface structure, a so-called 3D acquisition geometry is required. 3D acquisition geometry means that sources and receivers are usually distributed on a two-dimensional mesh on top of the acquisition surface, i. e., all three components of the location vector of sources $\vec{x}(S)$ and receivers $\vec{x}(R)$ can vary. However, in early days it was more common to use a so-called 2D acquisition geometry. Here, the sources and receivers are distributed along a straight line. After an appropriate rotation, the coordinates of the sources and the receivers vary only in the x_1 and x_3 component of their location vectors. From the acquisition geometry alone, the formulae for 3D wave propagation do not simplify. Furthermore, the 2.5D assumption for the underlying medium can simplify matters. Here, the x_2-component of the vector $\vec{x}$ remains constant and is usually equal zero. This implies that if the gradient operator is applied to the x_2-component, it will result in a zero as there are no changes in the x_2-direction.

With the assumption of a 2.5D medium, it can be sufficient to acquire the multicoverage dataset with a 2D acquisition geometry. Nevertheless, some considerations have to be taken into account:

the acquisition line should be parallel to the plane of symmetry. In this case, the observation plane coincides with the plane of symmetry and the 2.5D medium can be fully imaged by the resulting 2.5D model. However, if the acquisition line is not parallel to the plane of symmetry, then the observation plane is, in general, different for each simulated point within the 2.5D model. Thus, the orientation of the 2D acquisition line has to be parallel to the plane of symmetry. This can be achieved by additional information from earlier measurements and/or by geological informations.

From here on, I assume that a 2.5D medium is investigated with an acquisition line parallel to the plane of symmetry. Thus, the following considerations are carried out in the observation plane, i. e., the plane of symmetry.

3.2 Geometrical derivation of 2D ZO traveltime formula

At first, the CRS stack method depends on the redundancy in the data acquired in the frame of seismic reflection measurements. There, the sources and receivers are moved in such a way that each point on a reflecting interface in the subsurface is illuminated several times by source-receiver pairs with different offset, i. e., different distances between sources and receivers. In case of a 2D acquisition, the sources and receivers can be positioned along a usually straight line in many different configurations. It is common practice to acquire the data with a multitude of receivers. The receivers will record the seismic wavefield (or at least components of it) initiated by the time the source starts to emit energy to the subsurface, i. e., the source time which usually defines $t = 0$. These so-called shots are repeated many times from different locations along the 2D acquisition line. Thus, the location of sources and receivers can be described by scalar coordinates x_S and x_R, respectively. Additionally, the subsurface model is assumed a 2.5D medium (i. e., a 3D medium with no variations perpendicular to the 2D acquisition line in the x_1-x_2 plane). The 2D ZO CRS stack will also work for an arbitrary 3D medium but then the observation plane is in general different for each considered sample of the multicoverage dataset. The 2.5D medium assumption will not affect the kinematic aspects of the wave propagation considered in the subsequent thesis. Nevertheless, the dynamic aspects of wave propagation, i. e., the amplitude will be affected, but will not be further investigated here. In the following, the midpoint coordinate x_m and the half-offset h are used for the unique description of the recorded seismic traces instead of x_S and x_R. In the 2D case, these coordinates are defined as

$$x_m = \frac{x_R + x_S}{2} \qquad \text{and} \qquad h = \frac{x_R - x_S}{2} \qquad (3.1)$$

Then, the acquired data represent the multicoverage dataset and form a 3D data volume (see Figure 1.2) set up by the midpoint axis x_m, the half-offset axis h, and the traveltime axis t. The redundant information of one and the same point in depth is located along the so-called Common-Reflection-Point (CRP) trajectory which is given by a curve in the 3D (x_m, h, t) data volume.

3.2.1 CRP trajectory for homogeneous overburden

The approach of Höcht et al. (1999) for the derivation of a formula for a CRP trajectory is based on the principles of geometrical optics which are also used in the following. Firstly, the simplest case of one arbitrary reflector with a homogeneous overburden is considered. The layer above this reflector or interface has the constant velocity v and $z = 0$ is the measurement surface. The kinematic reflection response of the reflector is the traveltime surface $t = t(x_m, h)$ within the 3D multicoverage dataset. From this traveltime surface, the reflector can be fully recovered by computing the envelope of all isochrons defined by $t(x_m, h)$ and v. In case of a homogeneous overburden, the isochrons can be analytically expressed by

$$F(x, z; x_m, t(x_m, h)) = \frac{(x - x_m)^2}{\left(\frac{vt}{2}\right)^2} + \frac{z^2}{\left(\frac{vt}{2}\right)^2 - h^2} - 1 = 0. \tag{3.2}$$

This family of ellipses parameterized by x_m and h is directly deduced from Fermat's principle of stationary traveltime[1]. From the acquisition geometry, the multicoverage dataset usually contains redundant illuminations of many subsurface points. Thus, it is sufficient to consider only a subset of the family of isochrons: irrespectively of the source and receiver configurations, the envelope of the associated isochrons defines the illuminated part of the reflector. In the following, I consider the case of common-offset (CO) configurations, i. e., h = const. With the condition for the envelope

$$\frac{dF}{dx_m} = \frac{\partial F}{\partial x_m} + \frac{\partial F}{\partial t}\frac{\partial t}{\partial x_m} = 0, \tag{3.3}$$

the solution of equation (3.2) yields the envelope

$$x = x_m + \frac{t}{2h^2 t'}\left(\left(\frac{vt}{2}\right)^2 - h^2\right)\left(1 - \sqrt{1 + 4h^2\left(\frac{t\,t'}{t^2 - \left(\frac{2h}{v}\right)^2}\right)^2}\right), \tag{3.4a}$$

$$z = \sqrt{\left(\left(\frac{vt}{2}\right)^2 - h^2\right)\left(1 - \frac{(x - x_m)^2}{\left(\frac{vt}{2}\right)^2}\right)}, \tag{3.4b}$$

with $t' = \partial t/\partial x_m$. From equations (3.4), the solution for z is restricted to be real and positive. The simple case of ZO, i. e., $h = 0$ is not considered for this solution at the moment, but will be introduced later on as a simplification of the subsequent considerations.

Now, an arbitrary point $(\hat{x}_m, \hat{h}, \hat{t})$ lying on the reflection event is investigated. The associated traveltime $\hat{t}$ and its slope in midpoint direction $\hat{t}' = \partial t/\partial x_m|_{x_m=\hat{x}_m, h=\hat{h}}$ are both assumed to be known. The substitution of x_m, h, t, and t' with $\hat{x}_m$, $\hat{h}$, $\hat{t}$, and $\hat{t}'$ directly yields the corresponding reflection point $R = (\hat{x}, \hat{z})$ on the reflector. The tangent and—perpendicular to it—the normal to the reflector at point R are provided by calculating the first derivative of the CO isochron at this location.

Then, the intersection points of the tangent and the normal with the x-axis are calculated. The intersection points are denoted with x_T for the tangent and x_0 for the normal. From the results, the distance between both points without any further derivation reads

$$2r_T = x_0 - x_T = \frac{\hat{t}^2 - \left(\frac{2h}{v}\right)^2}{\hat{t}\hat{t}'}. \tag{3.5}$$

From the derivation of this distance, it is obvious that the distance only depends on the location of the reflection point R and the orientation of the reflector at R, i. e., the reflector dip at R. Therefore, r_T

---

[1]Fermat's principle of stationary traveltime in optics: the actual path between two points taken by a beam of light is the one which is traversed in the least time.

is constant for all reflection events associated with R and turns out to be a characteristic quantity of the considered reflection point R. A single point on the reflection event $(\hat{x}_m, \hat{h}, \hat{t})$ and its slope $\hat{t}'$ are sufficient to determine this characteristic.

The normal to the reflector at R represents the ZO ray which illuminates the reflector at this point. With the ZO traveltime t_0, the length of the ZO ray is given by $vt_0/2$. Then, the point on the measurement surface where the ZO ray emerges is $(x_0, 0)$. Any point within the traveltime surface that is associated with the same reflection point R has to satisfy equations (3.4) with $x = \hat{x}$ and $z = \hat{z}$. Incorporating all these equations, a relation between the considered arbitrary point $(\hat{x}_m, \hat{h}, \hat{t})$ and any other point (x_m, h, t) that is located on the reflection event stemming from the reflection point $R = (\hat{x}, \hat{z})$ can be derived:

$$x_m - r_T \sqrt{\left(\frac{h}{r_T}\right)^2 + 1} = \hat{x}_m - r_T \sqrt{\left(\frac{\hat{h}}{r_T}\right)^2 + 1}, \tag{3.6a}$$

$$\frac{t^2 - \left(\frac{2h}{v}\right)^2}{1 + \sqrt{\left(\frac{h}{r_T}\right)^2 + 1}} = \frac{\hat{t}^2 - \left(\frac{2\hat{h}}{v}\right)^2}{1 + \sqrt{\left(\frac{\hat{h}}{r_T}\right)^2 + 1}}, \tag{3.6b}$$

where r_T is derived from equation (3.5). Thus, all points that fulfill equations (3.5) and (3.6) set up the searched-for CRP trajectory. It becomes obvious from the above mentioned relations that the explicit knowledge of the reflection point coordinates are not required to describe the CRP trajectory.

To come back to the ZO case, a point in the ZO section, i.e., $\hat{x}_m = x_0, h = 0$, and $\hat{t} = t_0$, is chosen. For this point, equations (3.6) simplify to

$$x_m(h) = x_0 + r_T \left(\sqrt{\frac{h^2}{r_T^2} + 1} - 1 \right), \tag{3.7a}$$

$$t^2(h) = \frac{4h^2}{v^2} + \frac{t_0^2}{2} \left(\sqrt{\frac{h^2}{r_T^2} + 1} + 1 \right), \tag{3.7b}$$

with

$$2r_T = \frac{t_0}{t_0'}. \tag{3.7c}$$

As in this simple case, a homogeneous overburden is considered, the horizontal slowness p, which is also called the ray parameter, can be used to rewrite equation (3.7c) in an alternative form. Hereby, the emergence angle α for the ZO ray measured with respect to the surface normal is introduced. Additionally, the emergence angle α coincides in this special case with the local reflector dip measured with respect to the measurement surface $z = 0$. With the well known relation $p = t_0'/2 = \sin\alpha/v$, equation (3.7c) reads

$$2r_T = \frac{vt_0}{2\sin\alpha}. \tag{3.8}$$

3.2.2 CRP trajectory for inhomogeneous overburden

As shown in the previous section for the case of homogeneous overburden, an analytic description of the CRP trajectory is available through equations (3.7). Now, at least an approximate generalization

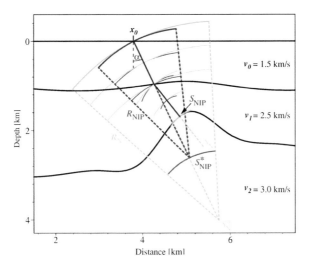

Figure 3.1: Simple model consisting of three homogeneous layers. An arbitrary ZO ray is shown in blue. The ZO ray impinges at S_{NIP} at the second reflector, i.e., the object point, and emerges at the surface at $(x_0, 0)$. S_{NIP}^* is the center of curvature for a NIP wavefront stemming from S_{NIP}. The center of the local curvature at S_{NIP} is denoted with S_N. The corresponding image point is shown as S_N^*. The same emergence angle α and radii of curvature R_{NIP} and R_N at x_0 would be obtained from a homogeneous model with the hypothetical reflector segment at S_{NIP}^* and velocity v_0 (taken from Mann, 2002).

in case of inhomogeneous media for the CRP trajectory will be derived in the following. Therefore, the concept of object and image points is used to establish a relation to the case of homogeneous overburden. This concept is well known in geometrical optics (see, e. g., Born and Wolf, 1959).

For illustration, the homogeneous overburden is now replaced by two homogeneous layers as shown in Figure 3.1. Here, the ZO ray emerges at the measurement surface at location $(x_0, 0)$ for an arbitrarily chosen reflection point S_{NIP} on the second interface. The suffix NIP indicates that the ZO ray impinges normally on the reflector at the so-called normal incidence point (NIP). The refraction of the ZO ray at the first interface is, of course, described by Snell's law.

To be able to relate properties of the reflector at each arbitrary reflection point S_{NIP} to quantities which can be observed at the measurement surface, two hypothetical eigenwave experiments have been introduced by Hubral (1983). The searched-for properties are the location, the local orientation and the local curvature of the reflector at each reflection point. The term "eigenwave" expresses that the respective wavefronts before and after the reflection at the point of interest are the same except for their direction of propagation. This is also further explained in Hubral (1983). In the following, I will only describe the eigenwave experiments by upgoing wavefronts. At first, a point source is placed at S_{NIP} which results in the so-called NIP wavefront. The change of the radius of wavefront curvature belonging to the NIP wavefront within the homogeneous layers is described by the transmission law for wavefront curvatures (Hubral and Krey, 1980)

$$R_{P_2} = R_{P_1} + v_i \Delta t, \tag{3.9}$$

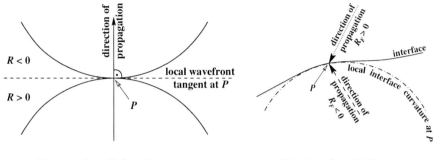

(a) propagating radii of wavefront curvatures   (b) local interface radii of curvatures

Figure 3.2: Sign convention for radii of curvature. (a) shows the radii of wavefront curvature R for a wavefront lagging behind its tangent at a considered point P. In this case, R is considered positive. If the wavefront is ahead of its tangent, R is considered negative. (b) The local interface radius of curvature $|R_F|$ at an arbitrary point P on the interface is shown as dash-dotted arc segment. For the dotted ray, the interface curvature appears convex. In this case, R_F is defined as positive. For the dashed ray, the interface appears concave which yields that R_F is considered negative. To summarize (a) and (b), all signs of radii of curvatures depend on the direction of propagation.

where P_1 and P_2 denote two points along the ZO ray within the actual considered homogeneous layer with the interval velocity v_i. Δt is the traveltime the wavefront needs to propagate from point P_1 to P_2. Thus, the transmission law states that the change of the wavefront curvature radius simply depends only on the length of the normal ray segment within the considered homogeneous layer. Such a behavior is usually called spherical divergence or, in a more general sense, geometrical spreading. The transmission law is only one component of the geometrical spreading. The geometrical spreading also accounts for changes of the radius of wavefront curvature due to refraction across interfaces. This additional change of the radius of wavefront curvature due to refraction across an interface is expressed by the so-called refraction law for wavefront curvatures which reads

$$\frac{1}{R_R} = \frac{v_R \cos^2 \gamma_I}{v_I R_I \cos^2 \gamma_R} + \frac{1}{R_F \cos^2 \gamma_R} \left(\frac{v_R}{v_I} \cos \gamma_I - \cos \gamma_R \right), \tag{3.10}$$

where $R_{I/R}$ stand for the radii of wavefront curvature, $v_{I/R}$ are the interval velocities, and $\gamma_{I/R}$ denote the angles between the direction of propagation and the normal to the interface. Hereby, the indices I (i. e., incident) and R (i. e., refracted) refer to the properties before and after the refraction, respectively. At the intersection of the ZO ray with the interface, the local interface radius of curvature is given by R_F. Figures 3.2 illustrate the sign convention used for the radii of curvatures in case of (a) propagating wavefronts and (b) local interface curvatures.

The third law related to the propagation of wavefronts is the reflection law. As I mentioned above that I will consider only the upgoing part of the eigenwaves, the reflection law is not required. Thus, please refer to Hubral and Krey (1980) for its description. Finally, the eigenwave wavefront emerges at the intersection of the ZO ray with the measurement surface at $(x_0, 0)$ after half the ZO reflection traveltime $t_0/2$. The radius of curvature at $(x_0, 0)$ is denoted R_{NIP} and α is the angle between the ZO ray and the surface normal, i.e., α describes the direction of propagation. If the NIP wavefront at $(x_0, 0)$ is considered for a homogeneous overburden with velocity v_0, the image point S^*_{NIP} is given by the center of curvature R_{NIP} (see Figure 3.1).

Now, the second eigenwave is considered in a similar way as before. However, the point source is replaced by an exploding reflector segment with the local curvature of the interface at S_{NIP}. The center of curvature for the considered reflector segment is depicted by S_{N} in Figure 3.1. As all rays are perpendicular to this reflector segment, the resulting wavefront is the so-called normal wavefront. This normal wavefront arrives also after the half ZO traveltime t_0 at $(x_0, 0)$ and with the same direction of propagation as the NIP wavefront. However, the radius of curvature R_{N} at $(x_0, 0)$ in general differs from R_{NIP} except for diffraction points which I will not consider at the moment. If again a circular approximation of the wavefront at $(x_0, 0)$ is used and a homogeneous overburden with velocity v_0, the center of curvature R_{N} defines the image point which is depicted as S^*_{N} in Figure 3.1. With the two eigenwave experiments, it is now possible to relate the properties of the reflector segment, i.e., its location, its orientation, and its curvature, to the ZO traveltime t_0, the emergence angle α, and the two radii of curvature R_{NIP} and R_{N}.

To end up with an approximation of the CRP trajectory for inhomogeneous overburden, the actual medium is replaced by an auxiliary medium with the constant velocity v_0. In case of iso-velocity layers as used in Figure 3.1, the constant velocity v_0 of the auxiliary medium coincides with the velocity of the first layer. In more realistic models, i.e., for arbitrary inhomogeneous models, v_0 is given by the near-surface velocity which has to be assumed to be constant at least in the vicinity of the emerging ZO ray at $(x_0, 0)$ and, so far, also to be explicitly known.

The wavefront curvatures and the emergence angle for the two eigenwave experiments are exactly the same in the auxiliary model if they stem from the hypothetical reflector segment depicted in magenta at the image point S^*_{NIP} as within the inhomogeneous model stemming from S_{NIP} (see Figure 3.1). In the auxiliary model, this reflector segment has the dip α and the radius of curvature $R_{\mathrm{N}} - R_{\mathrm{NIP}}$. Nevertheless, the ZO traveltime $t^*_0 = 2R_{\mathrm{NIP}}/v_0$ associated with the reflector segment in the auxiliary model in general differs from the ZO traveltime t_0 obtained in the actual, inhomogeneous model. Taking the circular approximations for the NIP and normal wavefronts into account, the traveltime difference is independent on the selection of the considered source/receiver pair. Thus, $t^* - t$ coincides with $t^*_0 - t_0$ which can be solved for t and reads

$$t = t^* - t^*_0 + t_0 = t^* - \frac{2R_{\mathrm{NIP}}}{v_0} + t_0 \,. \tag{3.11}$$

The above equation expresses the relationship between the reflection event obtained from the actual, inhomogeneous medium and the reflection event that stems from the hypothetical reflection point S^*_{NIP} within the auxiliary medium. From this relation, the CRP trajectory for inhomogeneous overburden can be approximated. Hereby, equation (3.7) yields the CRP trajectory for the hypothetical reflection point S^*_{NIP} with $t_0 = 2R_{\mathrm{NIP}}/v_0$ and $v = v_0$. Incorporating also equation (3.11) into equation (3.7) finally yields the approximate CRP trajectory for inhomogeneous overburden:

$$x_m(h) = x_0 + r_T \left(\sqrt{\frac{h^2}{r_T^2} + 1} - 1 \right),$$ (3.12a)

$$\left(t(h) - \left(t_0 - \frac{2R_{NIP}}{v_0} \right) \right)^2 = \frac{4h^2}{v_0^2} + \frac{2R_{NIP}^2}{v_0^2} \left(\sqrt{\frac{h^2}{r_T^2} + 1} + 1 \right),$$ (3.12b)

with

$$r_T = \frac{R_{NIP}}{2 \sin \alpha}.$$ (3.12c)

Please note that for equation (3.12a), Snell's law is satisfied in the auxiliary medium at the reflection point S_{NIP}^* rather than at S_{NIP} in the actual model. This introduces an additional approximation.

Now, not only a reflection point should be considered by its location and orientation but the kinematic reflection response of the entire reflector segment in the vicinity of the object point S_{NIP} should be taken into account. Assuming a circular normal wavefront, the approximation for the ZO case simply has to consider the radius of wavefront curvature at $(x_0, 0)$ and at an arbitrary point $(\tilde{x}_0, 0)$. Then, the traveltime difference $\Delta t = \tilde{t}_0 - t_0$ for the rays emerging at $(\tilde{x}_0, 0)$ and $(x_0, 0)$ is given by $2 \left(\tilde{R}_N - R_N \right) / v_0$. With the surface location $(x_v, 0)$ vertically above S_N^*, a triangle with the corner points at S_N^*, $(\tilde{x}_0, 0)$, and $(x_v, 0)$ is defined to express the following relation:

$$\tilde{R}_N^2 = (R_N \cos \alpha)^2 + (R_N \sin \alpha + (\tilde{x}_0 - x_0))^2.$$ (3.13)

Incorporating the relation between $\tilde{R}_N$ and $\tilde{t}_0 - t_0$, the above equation can be rewritten for $\tilde{t}_0$ and reads

$$\tilde{t}_0 = t_0 + \frac{2R_N}{v_0} \left(\sqrt{1 + \frac{(\tilde{x}_0 - x_0)^2}{R_N^2} + \frac{2 \sin \alpha (\tilde{x}_0 - x_0)}{R_N}} - 1 \right).$$ (3.14)

With the ray parameter, i.e., the slowness, $p = \sin \alpha / v$ where $\alpha = \tilde{\alpha}$ and $v = v_0$, the slope of the ZO hyperbola (3.14) defines the emergence angle $\tilde{\alpha}$ given by

$$\sin \tilde{\alpha}(\tilde{x}_0) = \frac{v d\tilde{t}_0}{2 d\tilde{x}_0} = \frac{\tilde{x}_0 - x_0 + R_N \sin \alpha}{R_N \sqrt{\frac{(\tilde{x}_0 - x_0)^2}{R_N^2} + \frac{2 \sin \alpha (\tilde{x}_0 - x_0)}{R_N} + 1}}.$$ (3.15)

So far, only the ZO reflection response of the hypothetical reflector segment and the CRP trajectory for the reflection point S_{NIP} have been derived. Now, the full multicoverage reflection response should be approximated which can be achieved by combining the derivations above. With this combined approach, a moveout surface in the multicoverage data volume can be set up. In principle, a CRP trajectory will be attached to each point given by the ZO reflection response. For a ZO location $\tilde{x}_0$, the CRP trajectory is given by equation (3.7) with the substitution $t_0 = \tilde{t}_0$, $x_0 = \tilde{x}_0$, $\alpha = \tilde{\alpha}$, and $R_{NIP} = \tilde{R}_{NIP}$. Hereby, $\tilde{t}_0$ and $\tilde{\alpha}$ are defined by equations (3.14) and (3.15) whereas $\tilde{R}_{NIP}$ remains undetermined for the moment. As the radius R_{NIP} is assumed to be constant along the normal wavefront, it changes in the same way as radius of the normal wavefront $\tilde{R}_N$:

$$\tilde{R}_{NIP}(\tilde{x}_0) = \tilde{R}_N(\tilde{x}_0) - R_N + R_{NIP} = \frac{v_0(\tilde{t}_0 - t_0)}{2} + R_{NIP}$$

$$= R_N \left(\sqrt{\frac{(\tilde{x}_0 - x_0)^2}{R_N^2} + \frac{2 \sin \alpha (\tilde{x}_0 - x_0)}{R_N} + 1} - 1 \right) + R_{NIP}.$$ (3.16)

With the time shift (3.11) which is constant for all CRP trajectories, equations (3.14) and (3.15) and the additional assumption that R_{NIP} is constant along the normal wavefront yield $\tilde{t}_0 - 2\tilde{R}_{\text{NIP}}/v_0 = t_0 - 2R_{\text{NIP}}/v_0$. Finally, the approximated CRP trajectory (3.12) for inhomogeneous overburden and equations (3.14)-(3.15) define a family of CRP trajectories parameterized by $\tilde{x}_0$:

$$x_m(\tilde{x}_0, h) = \tilde{x}_0 + \tilde{r}_T(\tilde{x}_0)\left(\sqrt{\frac{h^2}{\tilde{r}_T^2(\tilde{x}_0)} + 1} - 1\right), \tag{3.17a}$$

$$\left(t(\tilde{x}_0, h) - \left(t_0 - \frac{2R_{\text{NIP}}}{v_0}\right)\right)^2 = \frac{4h^2}{v_0^2} + \frac{2\tilde{R}_{\text{NIP}}^2(\tilde{x}_0)}{v_0^2}\left(\sqrt{\frac{h^2}{\tilde{r}_T^2(\tilde{x}_0)} + 1} + 1\right), \tag{3.17b}$$

with

$$\tilde{r}_T(\tilde{x}_0) = \frac{\tilde{R}_{\text{NIP}}(\tilde{x}_0)}{2\sin\alpha(\tilde{x}_0)}. \tag{3.17c}$$

The equations above represent the searched-for approximation of the kinematic reflection response that would be obtained from an arbitrarily curved reflector segment in depth. Nevertheless, the used parameterization with $\tilde{x}_0$ cannot handle caustics in the normal wavefront at the location $(x_0, 0)$. Höcht et al. (1999) described the searched-for approximation parameterized by $\tilde{\alpha}$ to be able to handle such caustics. As later on in the implementation of the CRS stack only Taylor expansions of equations (3.17) are used, it does not matter which representation is implemented. Thus, the reader is again referred to Höcht et al. (1999) for the description parameterized by $\tilde{\alpha}$.

As already mentioned, the parametric representation of the CRS response (3.17) will not further be used as its implementation is very difficult to handle. Much easier to evaluate for each contributing trace, is a function explicitly depending on the midpoint dislocation from the considered midpoint $x_m - x_0$ and the half offset h. Thus, Taylor expansions of the CRS response are used to obtain such explicit functions. The disadvantage of representations after the Taylor expansion is that the geometrical interpretation gets more complicated. Nevertheless, the second-order approximation is still preserved.

Two different second-order Taylor expansions can be used for the implementation of the CRS stack: one expansion for t and one for t^2. The first expansion is used in case of calculating a simple analytic approximation of the projected first Fresnel zone (for further details refer to Vieth, 2001; Mann, 2002). The second expansion serves as a representation of the CRS stacking operator within the coherence analyses and the stack itself. As mentioned at the end of Chapter 2, Ursin (1982) has stated that a hyperbolic traveltime surface suites better in most practical cases as an approximation of the true traveltime surface.

The Taylor expansion of t, which is also referred to as parabolic CRS approximation, is given by

$$t_{\text{par}}(x_m, h) = t_0 + \frac{2\sin\alpha\,(x_m - x_0)}{v_0} + \frac{\cos^2\alpha}{v_0}\left(\frac{(x_m - x_0)^2}{R_N} + \frac{h^2}{R_{\text{NIP}}}\right). \tag{3.18}$$

The above equation describes a paraboloid with vertical symmetry axis and the apex located in the ZO plane of the 3D data space (x_m, h, t). The apex is shifted in time and midpoint with respect to t_0 and x_0. Considering the case of a diffractor in the subsurface, the special case of a paraboloid of revolution ($R_N = R_{\text{NIP}}$) is obtained. In general, the paraboloid can be elliptic ($R_N R_{\text{NIP}} > 0$) or hyperbolic ($R_N R_{\text{NIP}} < 0$).

The second and more practical case of Taylor expansion for t^2 is the so-called hyperbolic CRS approximation. This surface has like the paraboloid its apex in the ZO plane of the 3D data space.

Furthermore, the same considerations with respect to the apex location apply for this approximation. Thus, the hyperbolic CRS approximation reads as follows:

$$t_{\text{hyp}}^2 (x_m, h) = \left(t_0 + \frac{2 \sin \alpha \, (x_m - x_0)}{v_0} \right)^2 + \frac{2 t_0 \cos^2 \alpha}{v_0} \left(\frac{(x_m - x_0)^2}{R_N} + \frac{h^2}{R_{\text{NIP}}} \right). \tag{3.19}$$

The above expressed quadric describes depending on the attributes a hyperboloid of two sheets. At first, a hyperboloid, an ellipsoid or a more general surface which can be elliptic in the ZO plane and hyperbolic in the CMP gather or vice versa.

3.3 2D ZO traveltime formula for rough top-surface topography

In the paragraph above, a planar measurement surface was required for the presented geometrical derivation of the CRS traveltime formula. However, the influence of topography on the recordings cannot be neglected in many cases of land seismic recordings. Thus, Zhang (2003) has introduced the handling of top-surface topography into the CRS stack. The fundamental steps of the derivation by means of ray theory have been presented in Chapter 2. There, the traveltime formula (2.112) for hyperbolic moveout has been derived and is here repeated again for convenience:

$$
\begin{aligned}
\tau (S', R')^2 = \;& \left(\tau (S, R) + \vec{x}^T (R') \, \vec{p}^{(x)} (R) - \vec{x}^T (S') \, \vec{p}^{(x)} (S) \right)^2 \\
&+ \tau (S, R) \, \vec{x}^T (R') \, \mathbf{M}^{(x)} (S, R) \, \vec{x}(R') \\
&+ \tau (S, R) \, \vec{x}^T (S') \, \mathbf{M}^{(x)} (R, S) \, \vec{x}(S') \\
&- 2 \tau (S, R) \, \vec{x}^T (S') \, \mathbf{R}^{(x)} (S, R) \, \vec{x}(R') \, .
\end{aligned}
$$

This equation is the most general representation of the traveltime surface to be used in the CRS stack. It is suited for processing 3D data sets which have been recorded in a 2D area on the acquisition surface and form a so-called 3D survey. For a 2D survey, i.e., recordings along an 1D line on the acquisition surface, the traveltime formula simplifies significantly. I just want to mention that the subsequent residual static correction (see Chapter 4) has been tested for such 2D surveys and, thus, only 2D processing results will be presented in Chapters 5 and 6.

Equation (2.112) already accounts for arbitrary top-surface topography by the third component of the location vectors which stands for the depth coordinate. To obtain the traveltime formula for planar measurement surfaces and to handle datasets without top-surface topography (e.g., for marine datasets), the depth coordinate has to be eliminated. This can be easily achieved by setting the third component equal zero, i.e., $x_3 (S') = x_3 (R') = 0$.

A third simplification to be mentioned is the case of a constant near-surface velocity. Thus, the velocity gradients in the vicinity of the source and receiver locations which are already considered in the above traveltime equation can be neglected and set to zero, i.e., $\vec{\nabla} v (S) = \vec{\nabla} v (R) = \vec{0}$. Unfortunately, the current implementation of the CRS stack does not account for such velocity variations at the moment.

In the following, the first and the last simplification mentioned above will be used to process 2D onshore datasets. There, top-surface topography is an often encountered and important problem which has to be considered in seismic exploration and during the applied seismic data processing. Before the traveltime formula (2.112) is implemented, some further assumptions are made:

- The azimuth angle θ is assumed equal zero which implies that the 2D survey has been carried out, without loss of generality, along a line in x_1-direction. In reality, the sources and receivers are not perfectly aligned especially not in the land data case (crooked line problem). I assume that the effect on the reflection traveltimes is small and can be neglected. Thus, the midpoints will be projected onto the best fit straight line for the processing with the CRS stack method.

- Furthermore, also all variables associated with the x_2-direction have to be constant. For convenience and to further simplify matters, this constant is usually zero.

- As the downgoing and upgoing wave along the normal ZO ray propagate in opposite directions, the rotation matrices coincide, i. e., $D(R) = D(S)$ and the slowness vectors have opposite signs, i. e., $\vec{p}^{(x)}(R) = -\vec{p}^{(x)}(S)$.

- The ray propagator matrix along the ZO central ray can be expressed by wavefront curvature of two hypothetical eigenwave experiments (as already described in the geometrical derivation above or as described by Hubral, 1983) and reads

$$\Pi(S,R) = \frac{1}{K_{\mathrm{NIP}} - K_{\mathrm{N}}} \begin{pmatrix} K_{\mathrm{NIP}} + K_{\mathrm{N}} & 2v_S \\ \frac{2K_{\mathrm{NIP}}K_{\mathrm{N}}}{v_S} & K_{\mathrm{NIP}} + K_{\mathrm{N}} \end{pmatrix}. \tag{3.20}$$

In the 2D case, the 2×2 wavefront curvature matrices of the hypothetical NIP-wave and normal wave reduce to scalars K_{NIP} and K_{N} in equation (3.20), respectively. Thus, also the ray propagator matrix reduces from a 4×4 to a 2×2 matrix. The relation of wavefront curvatures and their radii is simply given by $R_{\mathrm{NIP}} = 1/K_{\mathrm{NIP}}$ or $R_{\mathrm{N}} = 1/K_{\mathrm{N}}$. With the propagator matrix expressed in terms of the wavefront curvatures (3.20), the hyperbolic form of the traveltime surface can be rewritten as

$$\begin{aligned} \tau^2(S',R') = \ & \left(t_0 - \tfrac{2}{v_S}(\Delta m_{x_1}\sin\beta_S + \Delta m_{x_3}\cos\beta_S)\right)^2 \\ & + \tfrac{2t_0 K_{\mathrm{N}}}{v_S}(\Delta m_{x_1}\cos\beta_S - \Delta m_{x_3}\sin\beta_S)^2 \\ & + \tfrac{2t_0 K_{\mathrm{NIP}}}{v_S}(\Delta h_{x_1}\cos\beta_S - \Delta h_{x_3}\sin\beta_S)^2, \end{aligned} \tag{3.21}$$

where Δm_{x_1}, Δm_{x_3}, Δh_{x_1}, and Δh_{x_3} are components of the 2D midpoint and half-offset vectors:

$$\vec{m} = \begin{pmatrix} \Delta m_{x_1} \\ \Delta m_{x_2} = 0 \\ \Delta m_{x_3} \end{pmatrix}_{(3D)} = \begin{pmatrix} \Delta m_{x_1} \\ \Delta m_{x_3} \end{pmatrix}_{(2D)} = \frac{1}{2}\begin{pmatrix} x_1(R') + x_1(S') \\ x_3(R') + x_3(S') \end{pmatrix} \quad \text{and} \tag{3.22}$$

$$\vec{h} = \begin{pmatrix} \Delta h_{x_1} \\ \Delta h_{x_2} = 0 \\ \Delta h_{x_3} \end{pmatrix}_{(3D)} = \begin{pmatrix} \Delta h_{x_1} \\ \Delta h_{x_3} \end{pmatrix}_{(2D)} = \frac{1}{2}\begin{pmatrix} x_1(R') - x_1(S') \\ x_3(R') - x_3(S') \end{pmatrix}. \tag{3.23}$$

Furthermore, t_0 is the two-way traveltime along the ZO central ray. v_S is the near-surface velocity at the source point S which coincides in the considered ZO case with the receiver point R and should be a priori known. The geometrical information of the top-surface topography along the 2D acquisition line is incorporated into equation (3.21) by $\vec{m}$ and $\vec{h}$.

The three variables β_S, K_{NIP}, K_{N}, i. e., the three wavefield attributes, remain unknown in the parameterization of the CRS traveltime surface (3.21) and are to be searched for during the CRS stack. The current implementation of the CRS stack uses coherence analysis based on the semblance criteria (see Neidell and Taner, 1971; Mann, 2002) to obtain these so-called CRS attributes.

Finally, I want to emphasize that the presented traveltime surface (3.21) for the CRS stack with respect to a complex topography is given in a local Cartesian coordinate system. This is convenient for the handling of the complex topography. In case of a planar measurement surface as presented for the geometrical derivation above, a global coordinate system will be better suited. Due to the planar measurement surface which is usually also horizontal, the coordinates for the topography reduce to their x_1 component, i.e., $\Delta m_{x_3} = \Delta h_{x_3} = 0$. With $\beta_S = \alpha$, $K_{\text{NIP}} = 1/R_{\text{NIP}}$, and $K_N = 1/R_N$, equation (3.21) reduces to equation (3.19) (see Mann et al., 1999; Mann, 2002). For the cascaded approach of the CRS stack method considering rough top-surface topography, the CRS stack operator for smoothed topography is required. The traveltime formula used for this operator is a generalization of equation (3.19) and a special of equation (3.21) and reads in local Cartesian coordinates

$$
\begin{aligned}
\tau^2(S', R') = \; & \left(t_0 + \tfrac{2}{v_S} m_{x_1} \sin\beta_S\right)^2 \\
& + \tfrac{2t_0}{v_S}\left(K_N \cos^2\beta_S - K_0 \cos\beta_S\right) m_{x_1}^2 \\
& + \tfrac{2t_0}{v_S}\left(K_{\text{NIP}} \cos^2\beta_S - K_0 \cos\beta_S\right) h_{x_1}^2 \, .
\end{aligned}
\tag{3.24}
$$

For further details on how to obtain the smoothed CRS operator from the formula for rough top-surface topography, please refer to von Steht (2004).

3.4 NMO/DMO/stack vs. CRS stack

In this section, I will present a very brief comparison of conventional stacking methods like normal moveout/dip moveout/stack (NMO/DMO/stack) with the CRS stack, here for the case without topography. More detailed comparisons can be found in Müller (1999) and Jäger (1999).

Conventional processing chains are mostly analyzing CMP gathers, represented by the green planes in Figure 1.2, to obtain 2D velocity models of the subsurface. Many different velocity analysis methods (for further details see Yilmaz, 1987) are used to receive a stacking velocity. By varying this stacking velocity, a CMP traveltime curve fitting best to reflection events can be found along which the amplitudes are summed up, i.e., stacked. The aim of stacking is to improve the signal-to-noise (S/N) ratio by up to a theoretical factor of $\sqrt{N}$ (Yilmaz, 1987), where N is the number of contributing traces, e.g., in a CMP gather. In reality, this factor is smaller than $\sqrt{N}$: the signal is not always coherent, the fitted curve does not match exactly the true reflection traveltime curve, and noise (amplitudes caused by wind, traffic, ...) does not always interfere destructively during the stack.

The NMO/DMO/stack can be regarded as a so-called migration to ZO (MZO) but with some additional approximations. MZO is in principle exact for all dips, whereas the *small-dip* approximation (Hubral and Krey, 1980) approximation is used for the NMO/DMO/stack (for further details refer to Deregowski, 1986; Hale, 1991). Nevertheless, MZO and NMO/DMO/stack both aim to eliminate the offset dependency of reflection events in CMP and CO gathers in order to produce a stacked ZO section. Therefore, the MZO sums up all amplitudes along the CRP trajectories belonging to all reflection points on the ZO isochron defined by the two-way ZO traveltime t_0. The stacked signal is placed into point P_0 which is one point of the ZO section to be simulated. The fan-shaped MZO operator is illustrated in the upper part of Figure 3.3(a) for the case of one reflection point R with a constant velocity overburden (v_0). For this case, the ZO isochron is the lower semicircle with the center at x_0 and radius $v_0 t_0/2$. The MZO can be decomposed into three major steps: normal moveout corrections, dip moveout corrections, and the stacking.

Considering plane reflectors with dips, the dip moveout (DMO) tries to correct for the reflection point dispersal (see Figure 1.1(f)) occurring from the dip of the reflector. The traveltime for a single dipping

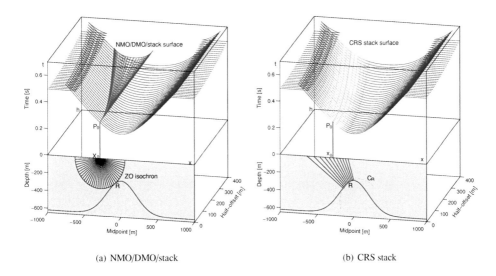

(a) NMO/DMO/stack   (b) CRS stack

Figure 3.3: Comparison of NMO/DMO/stack and CRS stack. (a) The fan-shaped NMO/DMO/stack operator is the reflection response of the ZO isochron. In the lower part of this figure, the ZO isochron is displayed a circle for this particular case of one constant velocity layer. (b) depicts the idealized CRS operator composed of all CRP trajectories stemming from the red arc segment at the reflection point R. This arc segment is the local interface curvature C_R.

reflector is (Levin, 1971)

$$t^2(h) = t_0^2 + \frac{4h^2 \cos^2 \delta}{v_{\mathrm{NMO}}^2} \qquad (3.25)$$

with the dip angle δ of the reflector. In the case of no reflector dip, i. e., $\delta = 0$, the reflection point dispersal vanishes and a common-depth-point (CDP) can be defined (see Figure 1.1(e)). The second term of equation (3.25) can be separated into a NMO and a DMO part,

$$t^2(h) = t_0^2 + \frac{4h^2}{v_{\mathrm{NMO}}^2} - \frac{4h^2 \sin^2 \delta}{v_{\mathrm{NMO}}^2} . \qquad (3.26)$$

The aim of MZO or NMO/DMO/stack is to provide a simulated ZO section with a high S/N ratio. With an improved S/N ratio, it is easier to identify reflection events because they more prevail the noise. The MZO operator is the fan-shaped surface shown in Figure 3.3(a) but it fits not very well to the true traveltime surface of the illustrated reflection event. Only along the light green line the operator sums up the amplitudes of the reflection event which is the CRP trajectory of R in the displayed case. The remaining part of the stack surface mainly adds noise to the stacked result. Thus, it deteriorates the stacking result because the noise does not always interfere destructively during the stack.

In contrast to the NMO/DMO/stack, the CRS stack makes use of the approximation for the CRP trajectories of the reflector segment in the vicinity of the reflection point R. Figure 3.3(b) shows the

45

idealized CRS operator which is build up by all CRP trajectories of the red arc segment around the considered reflection point R in the lower part. The fan-shaped stack operator of the NMO/DMO/stack, Figure 3.3(a), obviously takes a much smaller part of the multicoverage dataset into account than the CRS operator does. Therefore, the CRS stack has the potential to sum up more coherent energy of the reflection event which results in a high S/N ratio simulated ZO section.

Furthermore, the CRS stack method can also be considered a velocity analysis tool. The velocity model does not have to be explicitly provided for the stacking process as the velocity analysis is embedded in the CRS stack method. Only the near-surface velocity v_0 has to be known a priori as derived in the previous section that a geometrical interpretation and a CRS attributes-based redatuming is possible. Nevertheless, a reference velocity model can be used to constrain the search range which is useful in low-fold areas and to avoid the stacking of multiples to a certain extent. The hyperbolic CRS traveltime formula reduces for a CMP configuration to the form of the CMP stack formula with the stacking velocity v_{stack} which can be expressed in terms of CRS attributes:

$$t^2(h) = t_0^2 + \frac{4h^2}{v_{stack}^2} \qquad \text{with} \qquad v_{stack}^2 = \frac{2R_{NIP}v_0}{t_0 \cos^2 \alpha} \,. \tag{3.27}$$

Comparing this equation with equation (3.25) yields that the NMO velocity only depends on the CRS attribute R_{NIP}, i.e., $v_{NMO} = 2R_{NIP}v_0/t_0$. The NMO correction requires a velocity analysis to be performed in advance which is used to build up the velocity model for the subsequent NMO correction. This velocity analysis provides the stacking velocity instead of directly the NMO velocity and is usually applied at prominent locations only from which the rest is interpolated. In contrast, the CRS stack method performs a so-called high-density velocity analysis by searching for the CRS attributes at each sample within the ZO section that is considered. An additional advantage of this high-density velocity analysis is that the velocities change in such a way that almost no pulse stretch occurs (see Mann and Höcht, 2003).

Another advantage of the CRS stack is that it provides additional sections for each searched-for CRS attribute. The disadvantage of these sections is that the CRS attributes are meaningless in areas without detectable reflection events. Nevertheless, meaningful, reliable attributes can be identified due to their coherence which can be regarded as a kind of quality measure. The CRS attributes provide further information on the subsurface that can be used for subsequent processing steps. One application of the CRS attributes is a tomographic approach to build up a smooth macro-velocity model suitable for a subsequent post-stack migration or pre-stack migration (see Duveneck, 2004). This smooth velocity model is kinematically consistent with the data and well suited for ray tracing. Furthermore, the CRS attributes can enter into an event-consistent smoothing (see Mann and Duveneck, 2004) which can be regarded as a preconditioning. Hereby, the obtained CRS attributes and their coherence are used to align a smoothing window to the reflector dip which can be calculated from the emergence angle α. This smoothing reduces outliers and increases reflection event continuity in the simulated ZO section. Applying this smoothing to the CRS attributes is also important for the subsequently presented CRS-based residual static correction. On the one hand, the ZO traces are improved which will serve as pilot traces and, on the other hand, the outliers within the CRS attribute sections are reduced which stabilizes the moveout correction which is an essential step for the estimation of residual static corrections.

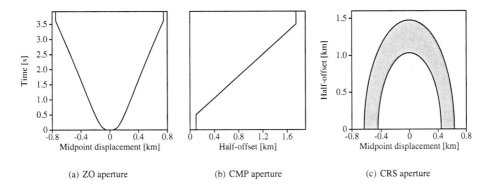

(a) ZO aperture   (b) CMP aperture   (c) CRS aperture

Figure 3.4: Slices of the CRS aperture. (a) shows an example of a common-offset slice, here at $h = 0$, i.e., the ZO aperture for a dominant frequency of 30 Hz and an average velocity ranging from 1.5 km/s to 3 km/s. The minimum ZO aperture is set to 50 m and the maximum to 750 m. (b) represents a common-midpoint slice, i. e., the CMP aperture. Between the points $(t_0, h) = (0.5 \text{ s}, 0.1 \text{ km})$ and $(3.5 \text{ s}, 1.75 \text{ km})$ this aperture increases linearly. Outside this area, the CMP aperture is extrapolated constantly. (c) is a time slice at $t_0 = 3$ s and illustrates the spatial extent of the CRS aperture. The shaded area is the part of the aperture that is tapered. The inner ellipse is usually 70% of the outer ellipse that basically covers the projected second Fresnel zone for ZO (taken from Mann, 2002).

3.5 CRS aperture

From Figure 3.3(b) which shows all CRP trajectories of the red arc segment, it is obvious that even this idealized CRS operator does not fit the true traveltime surface perfectly. Only for the red part of the local interface curvature, a sufficiently accurate approximation is achieved. Outside the sufficiently accurate approximation, the CRS operator would sum up noise which is not always interfering destructively. To avoid such an additional summation of noise, it is necessary to limit the summation range of the CRS operator. In the ZO-t-plane, the projected first Fresnel zone serves as a suitable measure to limit the ZO aperture (see Figure 3.4(a)).

In the ray theoretical ansatz, it is assumed that the seismic energy propagates along mathematical rays which have an infinitesimal volume. However in reality, the wave propagation is affected by a finite volume around the considered central ray as the recorded event is always a band-limited signal.

The first 2D Fresnel zone[2] is defined as a slice through the first 3D Fresnel volume around the central ray. This slice is perpendicular to the central ray and is selected to lie in the plane spanned by the central ray and the acquisition line. The first Fresnel zone is further defined as the area containing rays which signals interfere constructively at the receiver. For a mono-frequent signal with the period T, the first Fresnel zone along the t-axis would be the part with a maximum traveltime difference of $T/2$ to the traveltime of the central ray. This is the definition proposed by Kravtsov and Orlov (1990). The second Fresnel zone will contain rays which signals interfere destructively at the receiver and the

---

[2]Fresnel zone (after Sheriff, 2002): the portion of a reflector from which reflected energy can reach a detector within one-half wavelength of the first reflected energy.

third Fresnel zone has again constructive signals. Nevertheless, for a band-limited signal, only the first Fresnel zone is assumed to yield the major constructive contribution. If the central ray is normal to the reflector, the interface Fresnel zone coincides with the Fresnel zone perpendicular to the central ray as defined above. The projected (first) Fresnel zone for ZO, Hubral et al. (1993), is defined as the end points of the bundle of normal, paraxial rays stemming from the considered interface Fresnel zone. For further details and the calculation of the projected Fresnel zone from the CRS attributes, please refer to Vieth (2001) and Mann (2002).

The CMP aperture in half-offset-time planes (see Figure 3.4(b)) as well as the relation for the CRS aperture in half-offset-midpoint slice (see Figure 3.4(c)) can only be assigned empirically: the user has to define two points for the CMP aperture which serve as minimum and maximum aperture. In between these points, the CMP aperture is interpolated linearly. The CRS aperture is used with an elliptic form, where the half-axes are given by the ZO and the CMP aperture, respectively. The spatial extent of the CRS aperture is important for the later on described moveout correction which is a basic step of the CRS-based residual static correction (see Section 4.5). For further details on the implementation and some data examples, please refer to Mann (2002).

Chapter 4

Residual static correction

Before I can start to summarize the concept of residual static correction implemented into the CRS stack method, I will briefly explain the basic ideas and concepts of static corrections. Therefore, the definition of the weathering layer and the assumptions mandatory for the static correction will be explained in the following. Furthermore, some conventional residual static correction methods are briefly described and finally the CRS-based residual static correction will be explained.

4.1 The weathering layer

In most cases of land data acquisition, the area of interest is covered with a relatively thin layer of low seismic velocity. From the point of view of a geophysicist, this layer is called the weathering layer. The 'seismic' weathering layer is a near-surface low velocity layer in which the portion of air filled pore space of rocks is usually larger than of water filled (Cox, 1999). In contrast, for a geologist, the 'geological' weathering layer is the result of rock decomposition. Nevertheless, weathering is only one of several processes that can cause decomposition not only at the surface but also continuing downward. Mainly, the physical and chemical properties of gas, water, and organisms in and outside the rock are the driver for decomposition.

In general, the base and top of the weathering layer strongly fluctuate. This means that the elevation of the receivers is affected due to rapid changes in elevation and the thickness of the weathering layer does not remain constant. The thickness can vary between a few centimeters and up to 50 meters or even more. Furthermore, the weathering layer might expose a very heterogeneous velocity distribution in the vertical as well as in the lateral direction. This can be caused by the composition of sands and rock fragments at different stages of compaction and the differently filled pores. Compared with the geological weathering layer, the seismic weathering layer is usually thicker. The depth where the velocity changes to a significantly higher one or where the velocity stabilizes can be used to define the base of the weathering layer. In some cases, this base coincides with the water table.

The seismic weathering layer is often also called low-velocity layer (LVL). As a matter of fact, the effective or average velocity typically ranges between 500 m/s and 800 m/s which is relatively low compared to subweathering velocities of at least around 1500 m/s and higher. As the LVL usually consists of unconsolidated materials, its velocity strongly depends on water saturation and is related to compaction and thickness. Thus, the ratio of compressional and shear wave velocities in such media can vary between 1.3 and 10.0 . The reason for such a large deviation from the rule of thumb for

$v_P/v_S \approx \sqrt{3}$ is that the water saturation, or in a more general sense fluid saturation, has a strong influence on compressional wave velocities but nearly no effect on shear wave velocities. This is due to the fact that fluids have a usually negligible resistance against shear waves, i. e., the second Lamé parameter or shear modulus $\mu \approx 0$. The thickness of the seismic weathering layer is usually determined by refraction seismics. For the simplest case of a horizontal refractor with constant velocity layers, the principle of refraction seismics is explained in Appendix C. Another method commonly used to determine the LVL thickness is done by uphole surveys (for a more detailed description, please refer to Cox, 1999). This implies a sparse coverage of boreholes within the area of interest but at least a few boreholes should exist. The information provided by uphole surveys has to be interpolated to compensate the near-surface influence at each source and receiver location. To account for the geologic structure of the investigated area, this interpolation can be incorporated with reflection data, geologic data, and refraction data. But also a simple numerical interpolation is possible and sometimes adequate or at least more economic in practice.

Time-lapse surveys or monitoring can be influenced even stronger as some parameters associated with the near surface can vary with time. Some reasons for that are induced by the climate on a seasonal time base or even shorter, e. g., temperature changes, rainfall, tidal effects, ice movements, wind, recent erosion, deposition, earthquakes, and human activities to mention only a few. Thus, it is important to take these effects into account because time-lapse surveys or monitoring might be repeated after months or years and over several years.

The kind of topography has also an important influence on the characterization of the weathering layer (see Cox, 1999). The thickness of the weathering layer differs depending on whether there are sand dunes, a mountain front, youthful (characterized by active vertical erosion) or mature topography (no indication for variations in the near surface).

4.2 Assumptions for static correction

The time distortions due to all the facts mentioned above are mainly governed by the variations in thickness, elevation, and velocity of the seismic weathering layer. Thus, these deteriorations are the main cause for the loss of quality of land seismic data recordings. Figures 4.1 show some of the effects of (here: residual) statics. If the statics are not corrected for, the amplitude of the stacked result is not as high as theoretical possible, i. e., the gain due to stacking is less than $\sqrt{N}$ where N is the number of stacked traces. Also the frequency content decreases because of the misalignment of a peaks and the wavelet observed in the recorded traces is not preserved in the stacked traces. In this example, the small peak to the side of the maximum peak can be misinterpreted as a short period multiple or a ghost. Figure 4.2 shows the some usual attenuation functions of frequencies observed in the stacked result with respect to the extent of static shifts contained in the data.

To correct for these time distortions that in general degrade the stack quality, so-called static correction methods will be applied. However, three main assumptions have to be taken into account before the application of such static correction methods is justified:

- **Surface consistency**:
  The rays emerging at the measurement surface are assumed to propagate nearly vertical through the uppermost layer, i. e., the weathering layer. This is the most important assumption for static corrections. Figure 4.3 illustrates the difference of a surface consistent and a non-surface consistent velocity model. In the right part, the rays do not cross nearly vertical through the uppermost

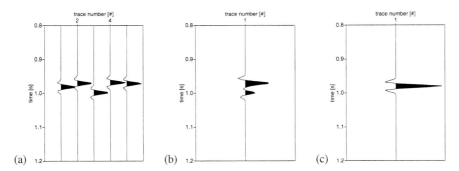

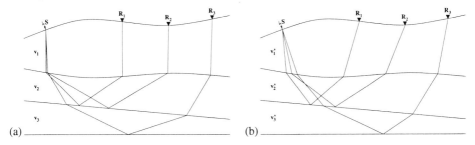

Figure 4.1: Example for residual static correction improvements. (a) shows some traces after an approximate NMO correction distorted by residual statics, (b) depicts the stack result of (a) without any residual static correction, i. e., stacked 'as is', and (c) is the stack result of (a) after residual static corrections have been applied.

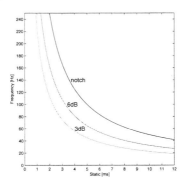

Figure 4.2: Frequency attenuation. The attenuation of frequencies in the stacked result is a function of the strength of static shifts contained in the dataset. Thus, static shifts act as a high-cut filter in the amplitude spectrum of the stacked result. Expressed in simpler words, the larger the statics the lower the frequency content.

Figure 4.3: Surface consistency and time invariance illustrated. For convenience, only few rays are shown. (a) shows an example of a model with LVL represented by velocity v_1 which fulfills the surface consistency and time invariance assumption. (b) depicts the same model as in (a) but with different velocities which now result in ray paths that are no more nearly vertical in the uppermost layer.

layer. Furthermore, the rays take significantly different paths through the weathering layer and, thus, will be influenced in different ways by the velocity inhomogeneities in the weathering layer. This will result in time shifts that are no unique property of the source or receiver location as it is mandatory for static corrections. In contrast to the non-surface consistent model, the velocity of the weathering layer is low compared to the deeper layers in the left part of Figure 4.3. Due to Snell's law (see equation (C.1)), the ray paths in the LVL are nearly vertical. Thus, the surface consistency is fulfilled and the time distortions caused by the velocity inhomogeneities become unique properties of the source and receiver locations, only.

- **Time invariance**:
  This assumption is strongly connected with the surface consistency. If the rays are nearly vertical in the LVL, also their associated traveltimes should be the same. As mentioned above, the influence of the weathering layer can change with time, but during a seismic measurement of usually up to 20 s, the effects of velocity inhomogeneities and thickness variations of the weathering layer are assumed to be invariant with time. Thus, the estimated time shifts should not depend on the traveltime of different reflection events. Figures 4.3 also illustrate that in (b) the ray paths are different and so the traveltime along the rays differ. In (a) the ray paths are surface consistent and the traveltime within the LVL does not depend on the depth of the reflector. This yields again that the time shifts become properties of the source and receiver locations and can be compensated with a unique static time shift.

- **Wavelet changes**:
  In addition to the surface consistence and the time invariance, it is necessary to assume that the weathering layer has no or at least the same effect on the wavelet of all emerging waves. If this is not the case, the recorded signal can have a positive maximum peak at one receiver and at an other receiver a negative maximum peak. On the one hand, this will influence the stacking of reflections events and on the other hand, the cross correlations for the estimation of statics can produce cycle skips due to phase shifts. The later presented implementation of residual static correction into the CRS stack method does not account for phase shifts.

4.3 Field static correction

The name field static correction is not connected with the velocity field in the weathering layer, it is more a historically based name as the field static correction was in early days applied during the field work by the field crew for a check of the data quality and a first interpretation. The field static correction is often also called datum or elevation static correction. The purpose of the datum static correction is to compensate the influence of the topography and the weathering layer. Therefore, usually a planar reference datum beneath the weathering layer is introduced. The sources and receivers are then shifted down or up to simulate the recorded signals of hypothetically conducted seismic experiments on the newly introduced reference datum plane (see Figure 4.4 (a)). To explain how the datum static correction is applied in practice, Figure 4.4 (b) shows a simple near-surface model with a single LVL.

The datum static correction is usually applied after the preprocessing (e. g., after deconvolution, trace balancing, etc.) and can be separated into two parts, firstly the weathering correction and secondly the elevation correction. The weathering correction tries to compensate for the effects mainly caused by

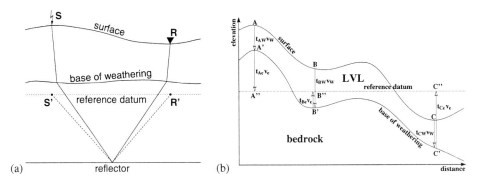

(a)   reflector   (b)   distance

Figure 4.4: Example for datum static correction. (a) shows the datum static correction of source S and receiver R down to a planar reference datum. Then, S' and R' are the source and receiver of a hypothetical measurement conducted on the introduced reference datum. (b) illustrates three different combinations of weathering and elevation corrections. Hereby, the weathering correction is used to eliminate the influence of the LVL and redatumes the traces to the floating datum of the base of the weathering layer. Then, the elevation correction is used to further redatum the traces from the floating datum to a planar reference datum.

the velocity inhomogeneities and thickness variations of the LVL. This means that after the weathering correction the new reference datum will be the base of the weathering layer which usually is a kind of floating datum. In Figure 4.4 (b), all points have to projected from the surface down to the base of the weathering layer by applying the weathering correction depicted as t_{AW} for point A to A', t_{BW} for point B to B', and t_{CW} for point C to C' calculated with the velocity v_w which is not necessarily the same for all locations. Thus, the traveltimes recorded at the surface have to be adjusted to obtain a new set of traveltimes that would have been recorded as if the data had actually been recorded at the base of the weathering layer. Here, the source type plays a small but important role. For vibrators used as sources, the weathering correction has to be applied for the sources as well as for the receivers as the sources are located right on top of the surface. For detonations, it is common to place the sources down in boreholes beneath the base of the weathering layer for a good energy coupling. Thus, the sources will not be influenced by the weathering layer and do not require a weathering correction.

Before proceeding with the elevation correction, a sign convention is necessary. Thus, correcting for positive elevations with respect to the planar reference datum corresponds to correct the traveltimes with negative time shifts which reduces the reflection traveltime. Now, the next step is to apply the elevation correction. The elevation correction is used to compensate the influence of the topography of the base of the weathering layer in the example of Figure 4.4 (b). Usually, the reference datum is a plane horizontal datum to which again every point has to be projected vertically. The purpose of the elevation correction is to finally get an easier to process and to interpret data set. If the reference datum is located beneath the base of the weathering layer, an error is introduced as the surface consistency holds only for the LVL. But if the reference datum is above the base of the weathering layer, an error should not occur as the elevation correction is performed along the rays are assumed to cross the LVL nearly vertically in compliance with the surface consistency. At the moment, errors depending on the reference datum level are neglected but will be discussed later. Thus, the traveltime t_{Ae} is calculated between points A' and A'' with the velocity v_e which should usually match the uppermost bedrock velocity and is chosen to be constant for all points. In analogy to point A', points B' and C' are similarly corrected. Finally, the full datum correction at point A can be written as $t_A = t_{AW} + t_{Ae}$. At points B and C, the signs have to be accommodated according to the relative location of the considered datum. This yields the datum correction for point B as $t_B = t_{BW} + t_{Be}$ and for point C as $t_C = t_{CW} + t_{Ce}$. Now, consider a source at point A and a receiver at point B. Then, the datum correction for this case according to the previous considerations is given by

$$t_{AB} = t_{AW} + t_{Ae} + t_{BW} + t_{Be} = -\frac{\overline{AA'}}{v_w} - \frac{\overline{A'A''}}{v_e} - \frac{\overline{BB'}}{v_w} + \frac{\overline{B'B''}}{v_e}. \tag{4.1}$$

To determine the required time shift for the datum correction t_{AB}, some parameters have to be known. The weathering correction depends on the elevation of source and receiver, the thickness, and the velocity of the LVL. The elevation correction depends as well as the weathering correction on the elevation of source and receiver, the thickness, and the velocity of the LVL and/or on the velocity between the base of the weathering and the considered planar reference datum. These parameters of the near surface are usually determined from refraction seismics (see Appendix C) and/or uphole surveys. As these parameters can change not only smoothly but also rapidly in vertical as well as horizontal direction, a precise knowledge of the near surface is crucial for an accurate correction. However, an accurate correction is usually inefficient in terms of time and costs.

Another important question is how to choose the depth of the reference datum? In general, two conditions should be satisfied: firstly, the datum surface should be a horizontal plane atop constant velocity layers, to reduce the complexity of the subsequent processing steps. Secondly, the static corrections should be kept as small as possible to introduce only small errors if the surface consistency is not fulfilled. In some areas, the first condition can only be achieved if the datum is chosen to be located at a significantly large depth below the measurement surface. This might be necessary if local topographic features are present within the measurement area. This feature (e. g., a very steep mountain) can cause an load effect which results in a slightly higher acoustic impedance beneath due to the increased pressure. In general, the so introduced change in acoustic impedance decreases with the distance from the highest magnitude of such a feature (see Widess, 1946). This suggests to preferably choose a deep reference datum. However, this preference stands in contradiction to the above mentioned second condition. The deeper the reference datum is chosen, the larger the required time shifts and the larger the misfit of the vertical projection to the true ray paths. The assumption of surface consistency is only fulfilled in the LVL and, thus, the deeper the reference datum the greater the in general not vertical ray path beneath the base of the weathering layer has to be considered for the datum static correction. A consequence of large field static corrections is that they violate the time consistency and, thus, the reflection traveltimes after the correction deviate from a hyperbolic relationship for the actual NMO velocity (see Profeta et al., 1995). This results in a deviation of the stacking velocities determined from NMO corrections to the correct ones. This error is called residual NMO (for further details, please refer to Cox, 1999).

A solution to keep the residual NMO minimal is the concept of an intermediate so-called floating datum. Here, the floating datum is chosen in such a way that it is close to the surface to obtain small field static corrections. There exist many different methods to map a trace to an intermediate datum, I will mention only one. Hereby, the mean of all field static corrections within each CMP gather is calculated but not applied. Then, each trace is corrected with the difference between the mean of the corresponding CMP gather and the actual field static correction. Afterwards, the NMO correction is applied. To finally correct the data to the reference datum, a further time correction with the mean

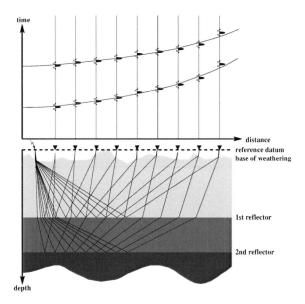

Figure 4.5: Example for residual statics. This figure illustrates the three main reasons for residual statics. In the lower part, the rapid changes in elevation as well as the thickness variations of the weathering layer are depicted. Velocity inhomogeneities are not indicated to reduce the complexity of this figure. In the upper part, two hyperbolic stacking operators are indicated by the red and blue curves. The residual statics become visible as the wavelets are shifted with respect to the stacking operators.

field static correction is performed. The advantage of this concept is that it avoids distortions of the near-surface velocities (which can be due to shifting shallow events to negative traveltimes) and also avoids deformation of the hyperbolic shape of near-surface reflections prior to NMO correction.

In general, field static corrections are usually applied to onshore datasets, only. In case of marine acquisitions, it is assumed that a LVL at the sea bottom does not exist and the water above the sea floor provides a planar measurement surface and has a more or less constant velocity. Nevertheless, in areas of rapid water-bottom topography or with a variable low-velocity material below the water-bottom, marine field static corrections are computed in analogy to the field static corrections for the more common land case.

4.4   Conventional residual static correction methods

As mentioned in the previous paragraph, the precise knowledge of the near surface is mandatory for an accurate datum static correction. If the near-surface parameters are not precise enough due to, e. g., a smooth description of the near surface, there might still remain small static time shifts that the datum correction did not compensate for. Additionally, small errors can be introduced due to rapid changes in elevation, strong velocity inhomogeneities, and small scale variations of the thickness of the weathering layer. Here, the residual static corrections come into play to further improve the image quality. Figure 4.5 illustrates the basic ideas of residual statics. In the lower part, the ray paths are depicted for the simple case of two horizontal reflectors. Here, the surface consistency is fulfilled as indicated by the nearly vertical ray segments in the LVL. The source and receivers are already corrected to the planar reference datum (dashed horizontal line), but due to the rapid changes in elevation, the thickness variations from receiver to receiver, and the velocity inhomogeneities of the LVL some time shifts remain in the data that are called residual statics. The residual statics are illustrated in the upper part where the red and blue curves stand for two stacking operators. It is obvious that the wavelets are shifted with respect to these stacking operators and will decrease the stacking amplitude. Furthermore, the frequency content is decreased and the wavelet is deformed as already indicated by Figures 4.1.

To achieve surface consistency, residual static correction techniques have to provide one exclusive time shift for each source and receiver location. It is most common to apply the residual static correction after the application of datum static corrections. Nevertheless, residual static corrections can also be applied before the datum corrections. However, this requires a more sophisticated stacking operator that is able to directly account for the topography. If the stacking operator does not account for the topographic undulations, the residual static correction is often unphysical and mainly provides 'cosmetic' improvements of the final stacked section.

The time shifts resulting from residual static correction analysis can be basically divided into several terms. The following definition implies that the traveltimes are moveout corrected before by, e. g., an approximate NMO correction. Then, the reflection events in each CMP gather are considered to be misaligned due to a source static, a receiver static, a residual moveout, and additional terms depending on the used method. The most commonly used terms (see Taner et al., 1974; Wiggins et al., 1976; Fitch, 1981; Cox, 1999) are

$$t_{i,j,h} = t_{r_i} + t_{s_j} + G_{k,h} + M_{k,h}X_{i,j}^2 \qquad \text{with} \qquad k = \frac{i+j}{2}, \tag{4.2}$$

where t_{r_i} and t_{s_j} are the receiver and source static at r_i, the receiver location of the ith receiver and at s_j, the source location of the jth source, respectively. $G_{k,h}$ is the structural term which is an arbitrary time shift for the kth CMP gather along the hth horizon and depends on the subsurface structure. If only traces within a CMP gather are considered for the cross correlations, then $G_{k,h}$ can be neglected as it remains constant within single CMP gathers. $M_{k,h}$ is the residual moveout term at the kth CMP gather again for the hth horizon. As this residual moveout is offset dependent, it is multiplied with the squared source-receiver distance, i. e., the squared offset $X_k^2 = \left(s_j - r_i\right)^2$. Considering the time invariance assumption, the index h can be omitted as it should not influence the resulting time shifts. There can be more terms incorporated into the calculation of residual time shifts, e. g., the cross-line dip coefficient $D_{k,h}$. Here, the cross-line dip depends on the considered reflector and modifies equation (4.2) in such a way that it reads

$$t_{i,j,h} = t_{r_i} + t_{s_j} + G_{k,h} + M_{k,h}X_k^2 + D_{k,h}Y_k \qquad \text{with} \qquad k = \frac{i+j}{2}, \tag{4.3}$$

where Y_k is the distance of the considered kth CMP to the reference line.

4.4.1 Linear traveltime inversion

After the field static corrections have been applied, to each trace of a dataset, recorded above a LVL, a total static time shift is assigned after equation (4.2), here without the dependency on the considered horizon, given by

$$t_{i,j} = t_{r_i} + t_{s_j} + G_k + M_k X_k^2 \quad \text{with} \quad k = \frac{i+j}{2}. \tag{4.4}$$

The number of different source locations is N_S and the number of different receiver locations is denoted as N_R. After calculating all CMP locations and an adequate CMP binning, N_G stands for the number of different CMP bins contained in the dataset. In general, the CMP fold is not constant. Usually, it decreases at the borders of the dataset depending on the used acquisition geometry. Furthermore, it might fluctuate around the possible maximum depending on the binning. For convenience, a constant CMP fold number N_F is used commonly represented by an average fold close to the maximum fold. With the definitions of these numbers, the number of equations for $t_{i,j}$ is given by $\approx N_G N_F$ and these equations contain $N_S + N_G + 2N_G$ unknowns. The last term for the total number of unknowns is due to the fact that the structural and the residual NMO terms in equation (4.4) are constant within a CMP gather. Typically, the number of equations is larger than the number of unknowns, i. e., $N_G N_F > N_S + N_G + 2N_G$. Thus, not all equations are independent of each other. "The linear simultaneous equations that define the static problem are said to be overspecified (there are more equations than unknowns) and underconstrained (they are deficient in the number of independent equations available to solve for the unknowns)." (after Wiggins et al., 1976).

An approximate solution can be obtained by least-squares methods. However, after the decomposition process, there is still a time difference $\varepsilon_{i,j}$ between each estimated total time shift $t'_{i,j}$ and the corresponding actual total time shift $t_{i,j}$. How to obtain $t_{i,j}$ necessary for the calculation of $\varepsilon_{i,j}$ will be explained later. The sum of the squared time differences variable that the least-squares methods need to minimize reads

$$Q = \sum_{i,j} \left(\varepsilon_{i,j} \right)^2 = \sum_{i,j} \left(t_{i,j} - t'_{i,j} \right)^2. \tag{4.5}$$

Inserting equation (4.4) for the estimated total time shift $t'_{i,j}$ leads to

$$Q = \sum_{i,j} \left(t_{i,j} - t'_{r_i} - t'_{s_j} - G'_k - M'_k X'^2_k \right)^2. \tag{4.6}$$

To obtain the least-squares solution, the minimum of Q has to be found. Therefore, all partial derivatives of equation (4.6) with respect to all variables have to vanish, i. e.,

$$\frac{\partial Q}{\partial t'_{r_i}} = \frac{\partial Q}{\partial t'_{s_j}} = \frac{\partial Q}{\partial G'_k} = \frac{\partial Q}{\partial M'_k} = 0. \tag{4.7}$$

From these conditions, a system of equations with $N_S + N_R + 2N_G$ equations and the same number of unknowns is obtained. This system of linear equations has now a unique solution. As mentioned above, the number of equations $N_G N_F$ can be reduced by the least-squares algorithm down to $N_S + N_R + 2N_G$ which is usually still a large number of equations. To finally end up with a solution of this system, there exist many different solution techniques. One of such methods is the iterative Gauss-Seidel method (see Wiggins et al., 1976). This method requires initial values for the unknowns. One set of initial values can be to set all unknowns equal zero, i. e., $t'_{r_i} = t'_{s_j} = G'_k = M'_k = 0$. The change of the unknowns after each iteration gives the stopping criteria for the iterations if it is below a specified threshold. Such a method is usually suited to estimate residual static corrections of up to half of the dominant period of the reflection events within the considered dataset.

As mentioned before, the actual total time shift $t_{i,j}$ is mandatory for the estimation of the residual static correction with the linear traveltime inversion. Most of the residual static correction methods

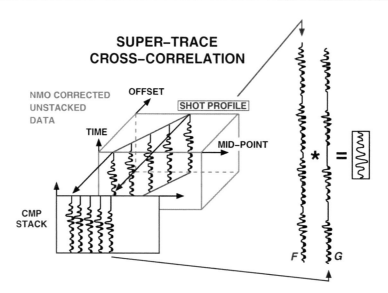

Figure 4.6: Example of super-traces for one moveout corrected shot profile. Super-trace F is concatenated from the traces of a shot profile. Super-trace G is built up of the corresponding CMP stacked traces. Afterwards, both super-traces are cross correlated to determine the corresponding source static. Figure taken from Ronen and Claerbout (1985).

are based on cross correlations. The cross correlations are, as well as the decomposition of the linear equation system, performed after the traces were moveout corrected. Otherwise, the offset dependency of the reflection traveltimes reduces the comparability of the traces with cross correlations to neighboring traces or even to no comparability if the offset increment between two neighboring traces is too large. Thus, I again want to emphasize that some kind of moveout correction has to be applied before the cross correlations should be used. For the cross correlations itself, two different approaches are available: firstly, all traces within the same CMP gather are cross correlated with each other. Secondly, the more common approach is to use a model or a so-called pilot trace. An example for a pilot trace can be simply the CMP stacked trace. To use a pilot trace with improved S/N ratio should provide an improved cross correlation result and, thus, a more reliable time shift. Hereby, it is assumed that the most reliable time shift for the residual static correction is associated with a maximum in the cross correlation result. The short-wavelength components of the source and receiver residual statics are assumed to have a mean of zero. Therefore, the remaining time shift estimated with the pilot trace approximates the structure provided the residual moveout component is small.

4.4.2 Stack power maximization

The linear traveltime inversion is sensitive to errors in the presence of ambiguities or noise within the cross correlation results from which the time shifts have to be estimated. To reduce this sensitivity, Ronen and Claerbout (1985) introduced their stack power maximization method. The basic idea of this approach is that the residual static correction aims to maximize the sum of the squared amplitudes of the stack, i.e., the power. The power of the stacked trace can be a quality indicator, if the power is maximized, all traces are aligned without any relative shift. In practice, each possible time shift Δt within reasonable limits can be applied to one single trace $F(t)$ within a CMP gather. Then, the resulting power $E(\Delta t)$ of $F(t)$ cross correlated with a pilot trace $G(t)$ is calculated as

$$E(\Delta t) = \sum_t (F(t - \Delta t) + G(t))^2 \overset{!}{=} \max \text{, or} \tag{4.8a}$$

$$= \sum_t \left(F^2(t - \Delta t) + G^2(t)\right) + 2 \sum_t (F(t - \Delta t) G(t)) . \tag{4.8b}$$

For a range of time shifts Δt, the first term of equation (4.8b) remains constant as it represents the sum of the total powers of trace $F(t - \Delta t)$ and $G(t)$. The second term is the cross correlation of both traces. Thus, the maximum of the cross correlation directly provides the maximum of the power and, associated with it, the desired time shift Δt. Now, consider the case that $F(t)$ is a trace of an individual source location and $G(t)$ is the corresponding stacked trace. The contribution of $F(t)$ should be excluded from $G(t)$ to avoid that the auto correlation of $F(t)$ contributes to the cross correlation result. Then, the time shift associated with a usually global maximum is an estimate of the residual source static correction as this time shift maximizes the stack power.

To additionally account for the subsurface structure, instead of cross correlating each individual trace from a source location and sum the cross correlations afterwards, Ronen and Claerbout (1985) introduced the so-called super-traces. Figure 4.6 illustrates how to build up the super-traces to estimate the time shift for an individual source location. At first, the dataset has to be moveout corrected, here the normal moveout correction was applied. Then, the traces contained within a shot profile are concatenated. In between the traces, zero segments are added to avoid cross correlating the wrong traces with each other. This is super-trace F shown in red. The second super-trace G is shown in blue and is built up as the concatenation of the corresponding CMP stacked traces, again with the same zero segments as for super-trace F. The cross correlation of both super-traces then provides the time shift of the considered source location. This is performed for all source locations. In analogy, the super-trace cross correlations are calculated with receiver profiles for all receiver locations. With the obtained source and receiver static corrections, the pre-stack dataset is corrected and the whole procedure can be applied again. Usually, the process is repeated 5-20 times to achieve a convergence in the results. Furthermore, some optional constraints can be accounted for within the procedure. These constraints aim to remove linear trends and drop outs from the estimated static time shifts and to find the searched-for global maximum of the stack power rather than a local one. For further details, please refer to Ronen and Claerbout (1985).

4.4.3 Non-linear traveltime inversion

As described for the linear traveltime inversion (see Section 4.4.1), the residual static corrections are estimated by linear inversion of traveltime deviations. Rothman (1985) has shown that the linearization of the difference of two static time shifts given by

$$\begin{aligned} \tau &= t_{i_1,j_1} - t_{i_2,j_2} \\ &= \left(t_{r_{i_1}} - t_{r_{i_2}}\right) + \left(t_{s_{j_1}} - t_{s_{j_2}}\right) + (G_{k_1} - G_{k_2}) + \left(M_{k_1} X_{k_1}^2 - M_{k_2} X_{k_2}^2\right) \\ &\qquad \text{with} \quad k_1 \frac{i_1+j_1}{2} \quad \text{and} \quad k_2 = \frac{i_2+j_2}{2} \end{aligned} \tag{4.9}$$

fails if τ contains errors due to large static time shifts and noisy data. Also the stack power maximization of Ronen and Claerbout (1985) runs into problems in such cases. A global optimization

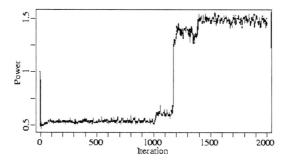

Figure 4.7: Simulated annealing technique. The Figure taken from Rothman (1986) depicts the normalized stack power with respect to the number of iterations. It is obvious that at first the stack power decreases due to the initial values of the static corrections. From the 100th iteration, the stack power begins to increase and finally reached an improvement of about 1.5.

technique can help to overcome these problems. Rothman (1986) introduced a method that makes use of probabilities instead of cross correlations. Hereby, probability distributions are used to draw random numbers required by a simulated annealing algorithm. These random estimates of static corrections are used to iteratively update the stack power until it converges to the maximum power (see equation (4.8a)). The simulated annealing technique is based on a Monte-Carlo technique to solve the global optimization problem. This method was introduced by Kirkpatrick et al. (1983) to solve the problem of the successful growth of crystals in chemistry. The successful growth in this case is tied to the global minimum of an optimization problem. Incorporating this method into the estimation of residual static correction, the method can escape from local maxima to find the global one which is achieved with the randomly chosen static corrections. An example of this technique is given in Figure 4.7. It shows the normalized stack power as function of the number of iterations performed. It is obvious that the stack power decreases at first due to the selected initial values. But from the 1000th iteration, the process starts to converge to the final solution which gives a gain factor of around 1.5 for the stack power in the end. This technique usually requires a lot of iterations which implies a high computational cost. Therefore, it should only be applied, if the data is expected to still contain large residual static time shifts. There exist other global optimization methods I can only mention here like, e. g., a genetic approach based on processes that mimic biological evolution (see Wilson et al., 1994; Cox, 1999).

4.5 CRS-based residual static correction

In the previous sections, I briefly reviewed some conventional residual static correction methods. Now, an approach that combines a residual static correction technique with the CRS stack method (described in Chapter 3) will be presented. This technique is in principle similar to the stack power maximization of Ronen and Claerbout (1985). Thus, the CRS-based residual static correction approach is, as well as most conventional methods, based on cross correlations and is a kind of linear traveltime inversion. So far, the residual static correction is implemented into the 2D ZO CRS stack method for plane or arbitrary top-surface topography, only. Therefore, only the 2D case is discussed in the following, but the strategy remains the same for 3D.

The 2D ZO CRS stack has proven over the last years that it can provide not only an improved S/N ratio in the simulated ZO section, but also additional information is provided by the so-called CRS attributes which represent the parameterization variables of the stacking operator. Some examples for such improvements in comparison to, e. g., the NMO/DMO/stack method have been published in Trappe et al. (2001). The additional sections of the CRS attributes are one section for the emergence angle α, one for the NIP wavefront radius of curvature R_{NIP}, and one more for the normal wavefront radius of curvature R_{N}. These attributes can be used to, e. g., determine a smooth macro-velocity model (see Duveneck, 2004) suited for a post-stack or pre-stack migration or to calculate a limited aperture for Kirchhoff pre- and post-stack migration.

The first step of the CRS-based residual static correction method is to perform at least the initial 2D ZO CRS stack to obtain the CRS attribute sections and the simulated ZO section. Each trace of the simulated ZO section serves as a pilot trace for the necessary cross correlations. Additionally, an event-consistent smoothing (see Mann and Duveneck, 2004) and a subsequent local optimization of the CRS attributes can be performed to further reduce the noisy character of the initial CRS attributes. The results of this so-called optimized 2D ZO CRS stack can also be used for the subsequent steps. However, this requires more processing time due to a local multi-parameter optimization of the attributes. The initial CRS stack differs from the optimized one by the strategy to obtain the attributes. The attributes of the initial search serve as starting values for the optimized search. Irrespectively of how the attributes are obtained, the CRS moveout correction is then realized with the previously obtained CRS attributes.

4.5.1 Moveout correction

As aforementioned for the conventional residual static correction methods, the moveout dependency of all traces which will be used for the cross correlations has to be reduced at least so far that the traveltimes can be assumed to be distorted only by residual statics. As derived in Chapter 3, the stacking operator (3.19) is given by the second-order traveltime approximation repeated here for a planar measurement surface:

$$t_{\mathrm{hyp}}^2 (x_m, h) = \left(t_0 + \frac{2 \sin \alpha \, (x_m - x_0)}{v_0} \right)^2 + \frac{2 t_0 \cos^2 \alpha}{v_0} \left(\frac{(x_m - x_0)^2}{R_{\mathrm{N}}} + \frac{h^2}{R_{\mathrm{NIP}}} \right).$$

with the ZO traveltime t_0, the near-surface velocity v_0, the emergence angle α of the ZO ray, the radius of curvature of the NIP wavefront R_{NIP} measured at x_0, and the radius of curvature of the normal wavefront R_{N} also measured at x_0.

To correct for the CRS moveout, the dependency on the half-offset h and the midpoint x_m in equation (3.19) has to be eliminated. Therefore, the CRS attributes of every time sample within the simulated ZO section are required. These attributes are provided by the initial or optimized search of the CRS stack method. With the knowledge of these attributes, the Common-Reflection-Surface can be transformed into a horizontal plane at time t_0 by subtracting the moveout given by

$$t_{\mathrm{moveout}}(x_m, h) = t_{\mathrm{hyp}}(x_m, h) - t_0 , \qquad (4.10)$$

where t_0 is simply given by the considered time sample of the simulated ZO section (see Figure 4.8).

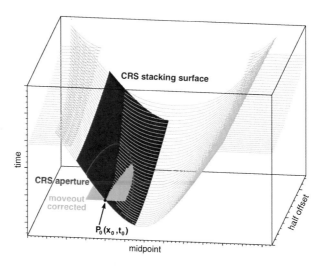

Figure 4.8: CRS moveout correction. The gray curves are CO curves of a dome-like structure in the subsurface plotted in the 3D data space. The front plane of this data space (midpoint, half-offset, time) is the ZO plane in which the simulated ZO section of the CRS stack is contained. The blue surface is an example of the CRS stacking surface for one time sample t_0 of the ZO section parameterized by α, R_{NIP}, and R_{N}. Here, t_0 is the ZO traveltime to which all traveltimes should be corrected to for the subsequent cross correlations. The green plane contains all the time samples within the CRS aperture after the moveout has been corrected, i. e., it is the part within the CRS aperture at a time slice through the data space at t_0.

This correction is performed for all t_0 given by each simulated ZO trace of the CRS stack. The result for one ZO trace is called CRS moveout corrected CRS super gather and contains all CRS moveout corrected pre-stack traces which lie inside the corresponding spatial CRS aperture (shown as red half-ellipse in Figure 4.8). Thus, the pre-stack traces are multiply contained in different CRS super gathers but with different moveout corrections in each super gather.

4.5.2 Cross correlation

Consider one CRS super gather, all traveltimes of each trace within this super gather are corrected to their corresponding ZO traveltimes t_0. Then, the obtained traces are assumed to be misaligned by the residual statics for the source and for the receiver locations. This has to be assumed to be consistent with the assumptions for static corrections at least concerning the surface consistency and the time invariance (see Section 4.2). To estimate the residual statics for all source and receiver locations within the whole dataset, cross correlations are performed between each moveout corrected trace and the corresponding trace of the simulated ZO section, i. e., the pilot trace. For one CRS super gather, there will be usually more than one cross correlation result for the subset of source and receiver locations belonging to the actual CRS super gather. How many cross correlation results will be obtained inside one CRS super gather for one source or receiver location depends on the midpoint and offset extension of the CRS aperture. Performing the cross correlations for all CRS super gathers within the dataset yields even more cross correlation results for each source and receiver location as the CRS super gathers will overlap again depending on the midpoint and offset extension of the CRS aperture. This is in principle similar to the super-trace cross correlation of Ronen and Claerbout (1985). However, the cross correlations are not performed with concatenated super-traces, but the cross correlation results will be summed up before the residual static corrections are estimated.

The main difference of the method by Ronen and Claerbout (1985) to CMP gather-based methods is that the correlation of the super-traces accounts for the subsurface structure because super-trace G of Figure 4.6 is a sequence of neighboring stacked traces and not of one stacked trace repeated multiple times. Super-trace F consists of all traces belonging to the same source or receiver location, respectively. The CRS stack accounts for the subsurface structure by means of the CRS attribute R_N which enters into the CRS moveout correction. R_N is the radius of curvature of the normal wave measured at the surface and can be associated with a hypothetical exploding reflector experiment.

To minimize effects due to different maximum amplitudes from trace to trace, the traces can be normalized before the moveout correction is performed. For this normalization, four choices in analogy to the coherence search of the CRS stack method are possible:

1. no normalization is applied, i. e., the original traces are moveout corrected and enter into the subsequent cross correlations.

2. instead of the original traces, the envelope is used. This is usually not beneficial for the subsequent cross correlations, as the envelope traces have a decreased frequency content compared to the original traces. This will make the estimation of the residual static corrections a bit harder. Furthermore, as the envelope contains only positive values, the cross correlations will be shifted that they also have no negative values. However, the assumption that the mean of both correlated traces is zero (which will be discussed later on) is violated.

3. the normalized traces are used. For the normalization, the original trace, i. e., the real part of the analytic signal will be divided by the envelope of the analytic signal. Thus, this normalization can be regarded as a kind of trace balancing as the modulus of the analytic signal can vary between -1 and $+1$ after the normalization. Nevertheless, the real part does not have to reach the maximum amplitude which implies that the imaginary part should be zero which is in general not the case.

4. the normalized analytic signal is used. As the analytic signal consists of a real and an imaginary part, both have to be moveout corrected and the subsequent cross correlations have to take the imaginary part into account, i. e., a complex cross correlation is required.

Before the cross correlations are performed, the moveout corrected traces can be additionally weighted. Here, the coherence section provided by the CRS stack can be used as such a weight. The coherence can be regarded as a kind of reliability measure for each sample of the ZO section. In most cases, semblance is used to calculate the coherence within a symmetric window around the stacking operator. During the search for the CRS attributes, many different stacking operators are tested while the set of CRS attributes with the maximum coherence is used for the subsequent stack. Applying this weight can provide a much clearer cross correlation result as unwanted noise is, in general, associated with a significantly smaller coherence value than contiguous reflection events.

Figures 4.9 illustrate the influence of noise on the result of the cross correlation. Hereby, Figures 4.9(a)-4.9(c) show the theoretical case of two traces with a frequency band-limited wavelet at

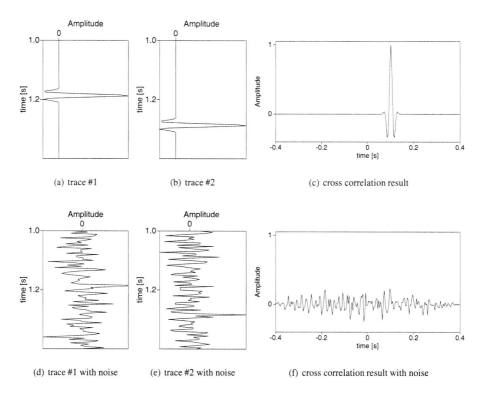

(a) trace #1   (b) trace #2   (c) cross correlation result

(d) trace #1 with noise   (e) trace #2 with noise   (f) cross correlation result with noise

Figure 4.9: Cross correlation without and with noise. If trace #1 and trace #2 are not influenced by ambient noise and are cross correlated against each other, then (c) depicts the theoretically possible result of the normalized cross correlation. The maximum at $\Delta t = +0.1$ ms of magnitude 1 indicates that the signals are identical. In practice, the traces are disturbed by noise, here with a S/N ratio of 2 shown in (d) and (e). The cross correlation result (f) is even more degraded by the noise as a reliable maximum is not visible any more.

different traveltimes without noise and the normalized result of cross correlating both traces. Thus, the maximum of the cross correlation is 1 with indicates that the signals are identical except for the time shift of $\Delta t = +100$ ms. For Figures 4.9(d)-4.9(f), artificial white noise was added with a S/N ratio of 2. Therefore, only the maximum of the wavelets of trace #1 and #2 is more or less observable. The shape of the wavelet can hardly be recognized. The cross correlation result of these noisy traces is fluctuating that much that a significant maximum with a sufficient reliability cannot be determined. However, the global maximum of the cross correlation is still at the correct time shift, most likely by chance in this special case of noise. The Figures 4.9 are intended to show extreme situations of noise levels. In practice, the result will hopefully be somewhere in between these two situations.

Another aspect concerning the cross correlations is the selection of the window length and position which defines the part of the traces to be cross correlated. Such a window should at least contain

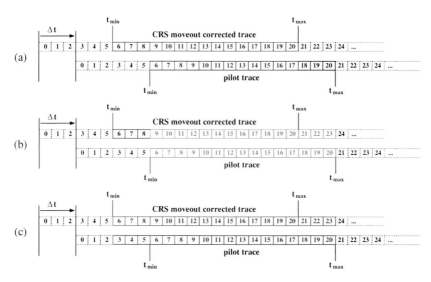

Figure 4.10: Windowing for cross correlations. (a) shows the standard windowing for the cross correlation of a CRS moveout corrected trace and a pilot trace. At a given time shift Δt, only the blue colored part will contribute to the cross correlation result as the traces are zeroed out outside the selected window given by t_{min} and t_{max}. (b) depicts the extended window for the CRS moveout corrected trace only and (c) illustrates the case for the extension of both traces, i. e., the CRS moveout corrected and the pilot trace.

---

one reflection event. The more reflection events contained, the smaller the discrepancy to the time invariance assumption. However in practice, it is common to cut off at least the first 50 ms, where the surface consistency is usually not fulfilled even for small offsets. Also for the CRS stack, the aperture at small traveltimes contains only few traces which decreases the reliability of the attributes. Thus, t_{min} is often chosen between 50 ms and 200 ms. t_{max} denotes the end of the cross correlation window and should be selected in such a way that on the one hand at least one primary reflection event is inside the window and on the other hand that not too much noise at large traveltimes contributes to the cross correlations. In practice, such a window has the length of about 1 s or more depending on whether there are visible reflection events at larger traveltimes or not. In standard cross correlation, the traces outside the selected window are zeroed out, see Figure 4.10(a). Thus, only a smaller part (here, depicted in blue) will contribute to the cross correlations the larger the considered time shift Δt. This can decrease the maximum theoretically possible result of the cross correlations. Thus, also the subsequent estimation of the residual static corrections can be affected. To account for such a possible decrease in the cross correlations, there are two choices to extend the correlation windows. On the one hand, only the CRS moveout corrected trace window can be extended to fully overlap with the constant window length of the pilot trace. This is illustrated by the green part in Figure 4.10(b) for a considered time shift Δt between both traces. This results in a constant number of contributing samples to the cross correlation and can be regarded as a pilot window moving over the moveout corrected trace. On the other hand, the windows of the moveout corrected and the pilot trace can be extended to fully overlap each other, see red part of Figure 4.10(c). The extension of both windows is used in cases of small correlation windows with large considered time shifts.

Last but not least, the cross correlations results can be normalized during their calculation. Therefore, the cross correlation of two stationary processes, i. e., traces $G(t)$ and $H(t)$, can be expressed (after Buttkus, 2000) as

$$\Phi_{GH}(\Delta t) = \lim_{t' \to \infty} \frac{1}{2t'} \int_{-t'}^{t'} \left(G(t') - \overline{G(t')} \right) \left(H(t' + \Delta t) - \overline{H(t')} \right) dt'$$

$$\text{with} \quad \overline{G(t')} = \lim_{t' \to \infty} \frac{1}{2t'} \int_{-t'}^{t'} G(t') dt' \quad \text{and} \quad \overline{H(t')} = \lim_{t' \to \infty} \frac{1}{2t'} \int_{-t'}^{t'} H(t') dt'. \tag{4.11}$$

Furthermore, if the traces can be assumed to behave like aperiodic functions, the forefactor $\frac{1}{2t'}$ in equation (4.11) is defined to be equal 1. Thus, the above equation is a more general representation of the cross correlation integral for continuous, infinite time series. For the estimation of residual static corrections, I assume that the mean of the traces is zero. Thus, the cross correlation simplifies and can also be expressed as a convolution:

$$\Phi_{GH}(\Delta t) = \int_{t'=-\infty}^{\infty} G(t') H(t' + \Delta t) dt' \equiv G(-\Delta t) * H(\Delta t), \tag{4.12}$$

where '$*$' denotes convolution. In practice, the signals are only recorded for a limited time and due to the above mentioned considerations, it is advantageous to perform the cross correlation only within a window. Furthermore, to reduce strong variations of the maximum of cross correlations of different traces, the cross correlations can be normalized. This normalization is given by dividing equation (4.11) with its cross signal energy. Together with the consideration that the $\overline{G(t')}$ and $\overline{H(t')}$ are zero, the cross correlation reads

$$\Phi_{GH}(\Delta t) = \frac{\int_{t'=t_{min}}^{t_{max}} G(t') H(t' + \Delta t) dt'}{\sqrt{\int_{t'=t_{min}}^{t_{max}} G^2(t') dt' \int_{t'=t_{min}}^{t_{max}} H^2(t') dt'}}. \tag{4.13}$$

As the traces are discrete time series, the integrals of equation (4.13) are transformed into summations of the samples within the considered standard or extended window. Thus, the discrete counterpart of equation (4.13) reads

$$\tilde{\Phi}_{GH}(\Delta t) = \frac{\sum_{t'=t_{min}}^{t_{max}} G(t') H(t' + \Delta t)}{\sqrt{\sum_{t'=t_{min}}^{t_{max}} G^2(t') \sum_{t'=t_{min}}^{t_{max}} H^2(t')}} \tag{4.14}$$

which is called discrete cross correlation and is widely used in various stages of data processing, e. g., vibroseis correlation or Wiener filtering. As already mentioned before, the correlation coefficients vary after normalization between +1 and −1. A +1 indicates that the two cross correlated traces are identical except for the time shift. The time shift is directly associated with the location of the considered extremum of the correlation result. In case of a −1, the traces are again identical except for the opposite sign of the amplitude and again the time shift. Thus, the higher the modulus of the correlation coefficient, the higher the similarity of both cross correlated traces.

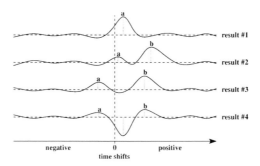

Figure 4.11: Possible correlation picks. In this example, four different correlation results are shown. The picks which yield the smallest time shifts are indicated with 'a', alternative picks with 'b', if they are 10 dB greater in magnitude than 'a'. Result #4 indicates a polarity reversed example where it is difficult to decide whether 'a' or 'b' will yield the better result. The minimum which belongs to the best correlation with the polarity reversed trace is not considered.

As mentioned before, the normalization of the analytic signal can be used for the subsequent cross correlations which implies complex cross correlation. Hereby, the discrete cross correlations reads

$$\Phi_{GH}(\Delta t) = \sum_{t'=t_{min}}^{t_{max}} G(t') H^*(t' + \Delta t) dt', \qquad (4.15)$$

where $H^*(t' + \Delta t)$ is the complex conjugate of $H(t' + \Delta t)$. For the subsequent estimation of the residual static corrections, the envelope of the complex cross correlation results will be used.

4.5.3 Estimation of residual static corrections

Due to noise (see Figures (4.9)), the maximum is usually not +1. Thus, different methods to estimate the "true" residual static correction are discussed in the following. All the considerations for improving the cross correlation results can be applied to each single cross correlation result. From the moveout correction, it is known that each source and receiver location is multiply contained in many CRS super gathers. All these cross correlations for common-source (CS) and for common-receiver (CR) locations are stacked. Assuming that the mean of all source and receiver static corrections is zero, the contributions of the receiver statics present in each source correlation stack should cancel each other and vice versa for each receiver correlation stack. Therefore, the decomposition of the total time shifts into source and receiver residual static corrections is possible.

From conventional methods, Figure 4.11 shows different possible correlation results and the different possible estimated time shifts. Result #1 is the most convenient case with one global maximum significantly higher than all other local maxima. It is obvious to take this global maximum as the estimated time shift. Result #2 has a local maximum closest to zero time shift, but also a 10 dB higher global maximum. Both maxima are possible, the local one is usually considered to keep the time shifts small. The stack power maximization of Ronen and Claerbout (1985) would rather take 'b' instead of 'a' as it is intended to maximize the stack power. Result #3 contains again two maxima but now with opposite signs. Last but not least, result #4 has a global minimum significantly higher than the two

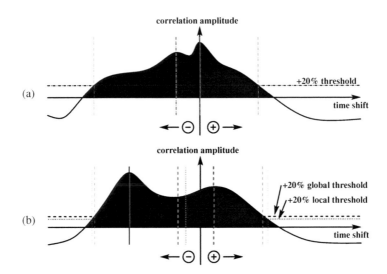

Figure 4.12: Examples for asymmetrical correlation stacks. Example (a) shows the case that the global maximum is at zero time shift (blue line) which results in the same time shift for the local maximum closest to zero time shift (dashed red line). Thus, both centers of the positive lobes around the global and the local maxima also coincide (dashed and dotted green lines). In this special case, the time shift estimated coincides most likely by chance with another local maximum. Example (b) depicts the situation that the global maximum and the local maximum yield time shifts with opposite sign. The center of the global and local maximum lobes yield slightly different estimated time shifts which both can be seen as a compromise for a further iteration to find the right time shift.

maxima. This indicates that one correlation trace fits better with a reversed polarity. Nevertheless, the global minimum will not be considered for the estimation of the time shift. Thus, one of the maxima is considered for the residual static correction. The complex cross correlation with the analytic signals of the moveout corrected and the pilot traces can help to fix the problem of reversed polarities. Here, the phase of the cross correlations is still affected by such reversals, but not the modulus. Thus, the modulus of the complex cross correlation should be evaluated to estimate the corresponding time shift.

The different methods to estimate the residual static corrections from the cross correlation stacks implemented into the CRS-based residual static correction method are:

1. **Global maximum**:
   Irrespectively of the resulting time shift, simply the global maximum is taken as the estimated residual static correction. See time shift indicated by the blue line in Figures 4.12.

2. **Local maximum closest to zero time shift**:
   Here, it is taken into account that residual static corrections should remain small. Thus, from all local maxima, the one closest to zero time shift is considered to be the estimated residual static correction. The corresponding time shift to this consideration is depicted by the dashed red line in Figures 4.12.

3. **Center of global maximum lobe**:

   The user can define a percentage of the global maximum to limit the positive area around the global maximum taken into account. Once the borders are defined, the center of this area is considered as the estimated time shift. This is used to overcome the fact that the autocorrelation possibly contained in the cross correlation stacks leads to many zero time shifts. Furthermore, assuming a symmetrical signal in the recorded traces should also yield a cross correlation result symmetrical to its global maximum. Figures 4.12 show examples of asymmetrical positive lobes around the global maximum. Here, the dashed green line indicate the considered time shifts.

4. **Center of local maximum lobe closest to zero time shift**:

   Similar to the method described before, a user-given percentage is applied but now to the local maximum closest to zero time shift. Again, the center between the borders given by the threshold is taken as the estimated time shift which is illustrated by the dotted green lines in Figures 4.12. Here in example (b), this method gives a compromise between the local and global maximum. This can help in a further iteration to find the optimum time shift.

As an outlook, another method for the extraction of residual static corrections from the cross correlation stacks can be to calculate the "center of mass" of the positive lobe either located around the global maximum or the local maximum closest to zero time shift. This calculation does not require thresholds to exclude small correlation coefficients due to noise as they only have small influence on the center of mass location. Furthermore, a "decreasing mass" to higher time shifts can be used to shift the center of mass closer to the assumption of small residual static corrections.

Figures 4.11 and 4.12 show only a few possible behaviors of the cross correlation stacks. There can be situations that are sometimes as difficult to interpret as shown in Figure 4.9(f). Some other effects can also influence the estimation like cycle skips or boundary effects. On the one hand, to reduce the likelihood of cycle skips the maximum time shift considered in the cross correlations can be limited. This should not be selected too small that at least one maximum remains recognizable in the cross correlation stacks to yield reliable residual static correction estimates. However, the smaller the maximum considered time shift, the faster the cross correlation calculation. On the other hand, to avoid boundary effects it is possible to define the minimum number of contributions to the cross correlation stacks. Due to the acquisition geometry, the number of contributions to the correlation stacks can decrease to only one contribution for certain source or receiver locations. Thus, if the number of contributions is too low at the boundary of the data set or due to acquisition gaps, then the estimated static might not be reliable and will be omitted for the subsequent correction.

Finally, the remaining estimated residual static corrections can be applied in several different ways of which I will only use two throughout this thesis. For both used methods, the corrections are directly applied to the pre-stack traces before the next iteration process is performed. The difference of the two applied methods is the search for the CRS attributes. The first iteration remains the same for both methods as the search for the CRS attributes is an essential part of the CRS stack method. But for further iteration, one can choose between a full iteration or a half iteration. A full iteration means: the search for the CRS attributes, the moveout correction, the cross correlations and stacks, the estimation of the time shifts, and the correction of the pre-stack traces. The other method is to assume that the search for the CRS attributes from the first iteration has provided the right attributes and, thus, the search step can be omitted. A third option, which will not be further discussed here, is to apply the corrections directly to the moveout corrected CRS super gather which, however, violates the assumptions of surface consistency and time invariance. If the static time shifts are applied to the moveout corrected CRS super gather, this means a dynamic time shift within the pre-stack traces because of the different moveout corrections. The latter two methods and combinations of all three methods have been investigated by Ewig (2003). In the end, I again want to emphasize that the CRS-based residual static correction is, as well as most conventional residual static correction methods, an iterative process. This will become more obvious in the following chapters with synthetic and real data examples.

Chapter 5

Synthetic data examples

The presented new approach for CRS-based residual static correction is tested on a simple full synthetic and a half synthetic data example. In this context, full synthetic means that the first data example (synthetic data example A) is generated from a simple given velocity model by ray tracing. Half synthetic means that the second data example (synthetic data example B) is a real dataset which is artificially distorted by random but surface-consistent time shifts. The used real dataset is a dataset with low complexity, i. e., the layers are more or less horizontally. In the following sections, I do not show all results as there are too many possible combinations of processing parameters, but only some of the most promising results for both data examples.

5.1 Synthetic data example A

The velocity model used for generating the multicoverage dataset consists of four well separated layers. In practice, the velocity contrast from the LVL to the first layer can be very high. Thus, most of the seismic energy is reflected at this first interface, i. e., reflectors beneath the base of weathering can hardly be seen in the seismic data. In this example, no LVL was modeled to simplify matters. This does not represent reality very well, but the LVL has been simulated by adding artificial but surface-consistent time shifts to the multicoverage dataset obtained by ray tracing. This has been intended to perform a first simple test to see whether the algorithm is able to correct for these time shifts or not. Thus, I will present only the results for one parameter set applied to this synthetic dataset. Later on, other parameter sets are compared for the second example which is closer to reality than this synthetic dataset.

5.1.1 Model and survey design

As mentioned above, this data example is fully generated from a given simple velocity model to test the CRS-based residual static correction method. For this purpose, the velocity model (see Figure 5.1(a)) consists of four constant-velocity layers, whereby the first reflector has been chosen to be horizontal, the second has a dip of $7°$, and the third has a syncline structure. The velocities are chosen as depicted in Figure 5.1(a) to mimic a situation encountered in land seismics. Nevertheless, a LVL was neglected in this model as the velocity contrast from the LVL to the first layer can be high and

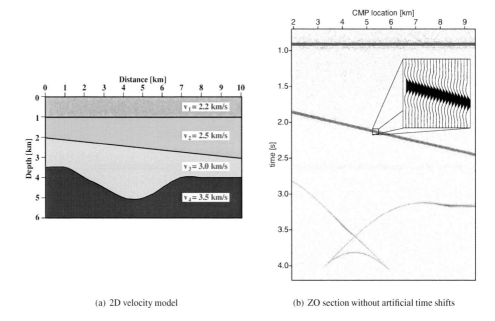

(a) 2D velocity model

(b) ZO section without artificial time shifts

Figure 5.1: 2D velocity model and CRS stack ZO section without artificial time shifts. (a) shows the 2D velocity model which consists of four constant velocity layers. The layers are separated by three interfaces, the first one is a plane horizontal interface, the second one is a plane dipping interface with a dip of $-7°$, and the last interface has a syncline structure. (b) depicts the simulated ZO section obtained from the CRS stack applied to the dataset with a S/N ratio of 3 before any artificial time shifts have been added.

---

only little energy would reach deeper reflectors. To further simplify matters, a top-surface topography is omitted. The velocity model depicted in Figure 5.1(a) represents the "true" subsurface structure as the aspect ratio is preserved. The time sections I present from here on, are not equally scaled as the downward axis is time and this axis can only be transformed to a depth axis with a varying velocity. Thus, the time sections are displayed with different scales which are selected to be convenient for comparison.

The 2D survey design is adapted from a marine data acquisition geometry. The first source point is located at 2000 m. The maximum offset between source and receiver is 2000 m and the minimum offset is 0 m. Thus, also the ZO section is contained in the generated multicoverage data set and can be used for comparison with the simulated ZO section of the CRS stack. The source and receiver spacings are both 25 m. With a streamer consisting of 81 receivers, the shots are carried out 321 times to reach the right border of the model at 10 km. Thus, 321 different source locations and 401 different receiver locations are spread along the 2D recording line from 0 km to 10 km and the pre-stack dataset consists of 26001 traces. The synthetic multicoverage dataset has been obtained by ray tracing and using a zero-phase Ricker-wavelet with a dominant frequency of 30 Hz, i.e., the maximum peak of

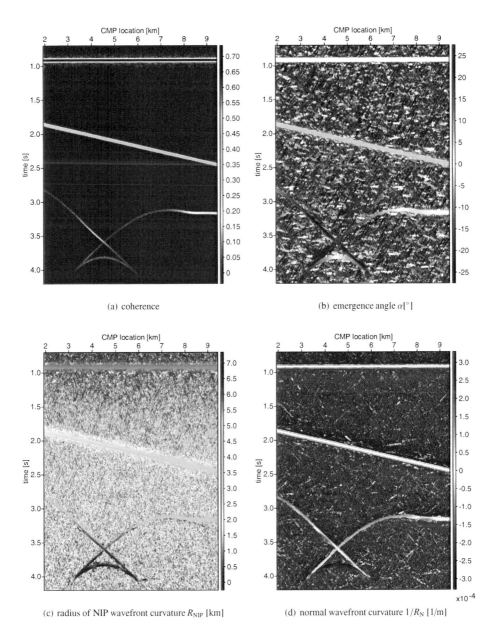

(a) coherence

(b) emergence angle $\alpha[°]$

(c) radius of NIP wavefront curvature R_{NIP} [km]

(d) normal wavefront curvature $1/R_{\mathrm{N}}$ [1/m]

Figure 5.2: Coherence and CRS attributes. (a) shows the coherence section belonging to Figure 5.1(b). (b), (c), and (d) display the CRS attributes emergence angle α, radius of NIP wavefront curvature R_{NIP}, and normal wavefront curvature $1/R_{\mathrm{N}}$, respectively.

Context	Processing parameter	Setting
General parameter	Dominant frequency	30 Hz
	Coherence measure	Semblance
	Data used for coherence analysis	Original traces
	Temporal width of coherence band	28 ms
Velocity and constraints	Near surface velocity	2200 m/s
	Tested stacking velocities	1500 ... 4000 m/s
Target zone	Simulated ZO traveltimes	0 ... 5 s
	Simulated temporal sampling interval	4 ms
	Number of simulated ZO traces	721
	Spacing of simulated ZO traces	12.5 m
Aperture and taper	Minimum ZO aperture	50 m @ 0.9 s
	Maximum ZO aperture	1000 m @ 2.5 s
	Minimum midpoint aperture	1000 m @ 0.9 s
	Maximum midpoint aperture	2000 m @ 2.5 s
	Relative taper size	30 %
Automatic CMP stack	Initial moveout increment for largest offset	4 ms
	Number of refinement iterations	3
Linear ZO stack	Tested emergence angles	-60 ... 60°
	Initial emergence angle increment	1°
	Number of refinement iterations	3
Hyperbolic ZO stack	Initial moveout increment for largest ZO distance	4 ms
	Number of refinement iterations	3
Hyperbolic CS / CR stack	Initial moveout increment for largest offset	4 ms
	Number of refinement iterations	3
Local optimization	Coherence threshold for smallest traveltime	0.01
	Coherence threshold for largest traveltime	0.005
	Maximum number of iterations	100
	Maximum relative deviation to stop	10^{-4}
	Initial variation of emergence angles	6°
	Initial variation of R_{NIP}	5%
	Initial variation of transformed R_{N}	6°
	Transformation radius of R_{N}	100 m

Table 5.1: Synthetic data example A: processing parameters used for the ZO simulation by means of the CRS stack. Some of the listed parameters are not relevant for the residual static correction, but are required for the reproducibility of the CRS processing. For further details, refer to Mann (2002).

the wavelet appears at the calculated arrival time. The sampling interval in time of the multicoverage dataset traces has been chosen as 4 ms.

Some of the parameters used for the CRS processing are listed in Table 5.1. Figure 5.1(b) shows the simulated ZO section obtained with the optimized CRS stack applied to the original dataset without artificial time shifts but with colored noise at a S/N ratio of 3. To obtain the colored noise, at first white noise has been added and at second a frequency filter has been applied. The frequency window

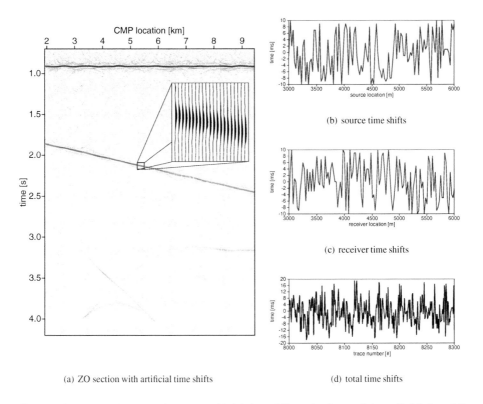

(a) ZO section with artificial time shifts

(b) source time shifts

(c) receiver time shifts

(d) total time shifts

Figure 5.3: CRS stack ZO section with artificial time shifts and subsets of the artificial time shifts. The simulated ZO section of the optimized CRS stack applied to the multicoverage dataset distorted with the artificial time shifts is shown in (a). (b), (c), and (d) depict only subsets of the artificial source, receiver, and total time shifts that have been added to the multicoverage dataset, respectively.

is chosen with the corner frequencies at 0 Hz, 10 Hz, 50 Hz, and 60 Hz that the frequencies between 10 Hz and 50 Hz are preserved. The stacked section is only shown between CMP #75 and #675 to get rid of less reliable results at the borders of the acquisition line due to a lower number of contributing traces, i.e., a lower redundancy in the multicoverage dataset. This lack of information is mainly caused by the acquisition geometry of a source moving from the left to the right at the surface with receivers only on the left of the source. The traces are only shown between 0.7 s and 4.2 s as the rest of the traces contains only noise. The small wiggle plot inside Figure 5.1(b) is a zoom of the second reflection event and depicts the wavelet after the stack which has the same shape as in the pre-stack traces. The comparison to the ZO section contained in the multicoverage dataset is omitted due to the low S/N ratio. With this noise, only the first reflection event would be recognizable in the provided ZO section. The noise will prevail the more over the reflection event amplitudes the larger the two-way traveltime and the more interfaces are traversed.

Figures 5.2 depict the coherence and the kinematic wavefield attribute sections obtained with the CRS stack method. These sections are the result of the local optimization applied to the event-consistently smoothed initial CRS attributes. The coherence of the optimized CRS stack is shown in Figure 5.2(a). Here, the three reflection events can be clearly recognized as their coherence differs from the coherence of the surrounding noise. The decrease of the coherence with larger two-way traveltime can have two reasons. On the one hand, the CRS aperture increases and, thus, more noise is incorporated in the semblance calculation. On the other hand, and this is the main reason for the decrease of the coherence, the noise level is adapted to the highest amplitude in the dataset without noise. In this case, the S/N ratio of 3 applies to the first reflection event and the S/N ratio is lower for the later reflection events. Figure 5.2(b) displays the section of the emergence angle α. As the first reflection event is a horizontal plane reflector, the emergence angle of each ZO ray is zero. The emergence angle along the second reflection event should be smaller than the true reflector dip as the ZO rays have to be refracted at the first reflector. With Snell's law due to the refraction at the first reflector, the emergence angle calculates to 6.16°. In case of an overburden with only horizontal or dipping plane reflectors, it is expected that the emergence angle for the second reflection event remains constant which is the case as shown in Figure 5.2(b). Along the crossing branches for the triplication of the third reflection event, it becomes obvious that no conflicting dip handling was applied. A detailed description of the detection of conflicting dips and its implementation can be found in Mann (2002). The radius of the NIP wavefront curvature and the curvature of the normal wavefront are shown in Figures 5.2(c) and 5.2(d), respectively. The stacking velocity changes along the wavelet, whereas the CRS attribute R_{NIP} is virtually constant along the wavelet. As a consequence, a pulse stretch is negligible (see Mann and Höcht, 2003). From the CRS attribute R_N, the local reflector curvature can be calculated if the overburden is known.

Even with the low S/N ratio of 3, the CRS stack was able to find the coherent reflection events. Now, the next step for this first test was to add artificial but surface-consistent time shifts for the source and the receiver locations. The surface-consistent time shifts are separately generated for source and receiver locations by random numbers varying between -10 ms and +10 ms. Thus, the total time shifts applied can vary between -20 ms and +20 ms for each trace of the multicoverage dataset. Figures 5.3(b)-5.3(d) display only subsets of the entire time shift sets applied as the entire multicoverage dataset contains 321 different source locations and 401 different receiver locations. With this new multicoverage dataset, again the CRS stack is performed. The resulting ZO section of the optimized CRS stack is shown in Figure 5.3(a). The processing parameters have been the same as for the CRS stack applied to the provided dataset without artificial time shifts. These processing parameters remained unchanged during the whole application of the CRS-based residual static correction, i. e., each iteration of the CRS stack has been performed with the same parameter set.

The ZO section in Figure 5.3(a) again contains a small zoom of the same area already shown in Figure 5.1(b). In comparison to the zoom corresponding to the ZO section from the original dataset, the zoom after the pre-stack traces have been shifted hardly shows the true shape of the wavelet. This is due to the applied time shifts which result in a blurred wavelet as already indicated by Figures 4.1. Furthermore, also the amplitude of the stacked result has decreased as an effect of the misaligned maxima of the reflection events.

5.1.2 Residual static correction

So far, only the multicoverage dataset has been distorted with the artificial time shifts and the CRS search for the attributes and the subsequent stack have been applied. Consequently, the CRS-based

Context	Processing parameter	Setting
Input	Data used for moveout correction	Original traces
	Moveout correction performed by	Optimized CRS attributes
	ZO section used as pilot traces	Fresnel stack
Cross correlation	Maximum correlation shift	40 ms
	ZO traveltime used for correlation	0.8 ... 4.1 s
	Correlation weight	CRS semblance
	Minimum number of 'live' samples per traces	0
Estimation of static correction	Data used for static correction estimation	Original correlation results
	Method applied	Center of positive area around local maximum closest to zero time shift
	Minimum threshold for method applied	30%
	Minimum number of contributions	sources: 0, receivers: 0

Table 5.2: Synthetic data example A: processing parameters used for the CRS-based residual static correction.

residual static correction is now performed to see whether it can correct for the surface-consistent time shifts. Table 5.2 summarizes the processing parameters provided for all iterations.

The minimum and maximum ZO traveltimes which limit the part of the traces considered for the cross correlations are chosen such as to contain all three reflection events. This is done in compliance with the assumption of surface consistency and time invariance as already mentioned in the previous chapter. Furthermore, the maximum considered time shift for the cross correlations is selected twice as large as the theoretically possible values to see if there are neighboring maxima in the cross correlation stacks and how large they are compared with the global maximum. Nevertheless, the maximum time shift remains small enough such that cycle skips are avoided in view of the strong separation of the reflection events.

Before the cross correlations can be performed, the moveout correction has to be applied. The effect of the so-called reflection event flattening can be observed for the example in Figure 5.4(b). Figure 5.4(a) shows the distribution of the traces in the x_m-h plane of the CRS super gather at CMP #240. This represents the shape of the maximal possible CRS aperture (compare with the green plane in Figure 4.8). As this example contains 3460 traces, only some selected CMP gathers (green and blue dots) are displayed in Figure 5.4(b). The arrows indicate the two-way ZO traveltimes at which the flattened reflection events should be observed. Due to the low S/N ratio, this is only recognizable for the first reflection event. At least, the linear increase of the CMP aperture (compare with Figure 3.4(b)) and the tapering effect (compare with Figure 3.4(c)) are clearly visible in this example.

Now, the cross correlations can be calculated between the pre-stack traces moveout corrected with the optimized CRS attributes and the pilot traces obtained from the optimized Fresnel stack. Optimized means that the stack is performed with the optimized CRS attributes and Fresnel means that if the user-given aperture ranges are too large, the CRS aperture is shrunk to the projected Fresnel zone. As already mentioned in Section 4.5, the moveout-corrected pre-stack traces are multiply contained in the overlapping CRS super gathers but with different moveout corrections. Thus, the correlations can be stacked for common-source (CS) and common-receiver (CR) locations at first for the considered CRS

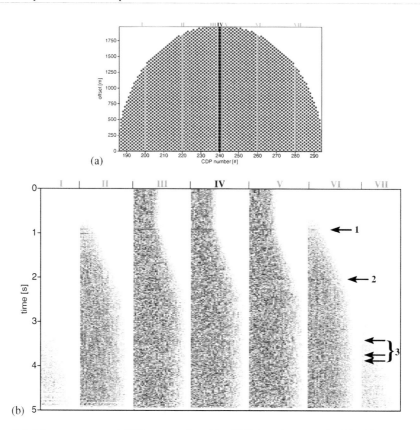

(a)

(b)

Figure 5.4: Moveout corrected CRS super gather. The CMP locations contained in this example for a moveout corrected CRS super gather are shown in the time slice (a) at $t_0 = 5$ s which represents the maximum CRS aperture for CMP #240. This super gather contains 3460 traces. Thus, only selected CMP locations (green and blue dots) are displayed in (b). The arrows indicate the traveltimes where the flattened reflection events should appear which are hardly recognizable due to the low S/N ratio of 3. For the third reflection event, three branches of the triplication should be visible.

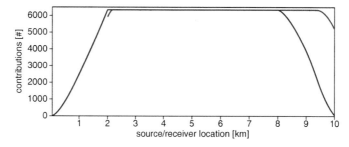

Figure 5.5: Number of contributions to correlation stacks. The red line shows the number of contributions for each source correlation stack and the blue line shows it for each receiver correlation stack.

super gather and at second as long as traces that contribute to the same source/receiver location are contained in other CRS super gathers. Figure 5.5 shows the number of cross correlations contributing to each source (red curve) and receiver location (blue curve). These curves depend, on the one hand, on the coverage of each CMP gather, where also the offset distribution of this coverage is important, and, on the other hand, on the spatial extent of the CRS aperture.

Then, the cross correlation stacks are used to estimate the source or receiver static correction usually associated with the time shift corresponding to a maximum in the cross correlation stacks. As I will show with examples later on, the global maximum is not always the best estimate. Thus, I applied the method explained as last method in Section 4.5.3. Hereby, the threshold of 30% of the local maximum closest to zero time shift defines the borders of the positive lobe around the considered maximum. The time shift associated with the center of this area is taken as the estimated static correction. The obtained source and receiver static corrections estimates are displayed in Figure 5.6 after each iteration applied to this synthetic example. Figures 5.6(a) and 5.6(b) show the source and receiver estimates after the first iteration. The second iteration results are depicted in Figures 5.6(c) and 5.6(d) and after the third iteration in Figures 5.6(e) and 5.6(f).

The results for the estimated static corrections (green curves) after the first iteration already fit quite well the artificially added ones (red and blue curves) except for some source or receiver locations. There the misfit is significant, e. g., at source locations at about 3500 m or 5500 m and receiver locations at around 4000 m and 5200 m). To refine the estimates of the first iteration and to reduce this misfit, a second and third iteration have been applied. In this case, the second and third iterations could refine the shape of the static curve but not the absolute values. Here, a systematic misfit mainly between -1 ms and -2 ms has been introduced by each iteration applied after the first iteration. This can be explained by looking at some of the cross correlation stacks itself shown in Figure 5.7.

The cross correlation stacks for a small subset of the different source (see Figure 5.7(a)) and receiver locations (see Figure 5.7(b)) show almost the same behavior. The low-frequency positive lobes are quite well centered around the curve of the artificial source (red curve) and receiver (blue curve) time shifts. However, there are high-frequency peaks positive to the one side of the global maximum and negative to the other side. These peaks in combination with the selected estimation algorithm for the determination of the static corrections are the main reason for the misfits after the first iteration and the systematic shift for further iterations. Here, on one side of the considered maximum, the minimum threshold is matched at a larger distance of the maximum than on the other side, where the minimum threshold is matched at a smaller distance of the maximum. Thus, the center which then defines the estimated static correction is shifted. To avoid the systematical misfit, the second and third iterations could have been performed with a different parameter set, where the method for estimating the static corrections has been changed to directly take the time shift associated with the global maximum. Nevertheless, the results will not be shown here as this dataset is only a first test.

As the estimated static corrections have been compared with the artificially added ones and showed an acceptable match, it is expected that also the simulated ZO sections after the third iteration of the CRS-based residual static correction has been recovered to match the section before the artificial time shifts were added. Thus, I present both sections side by side for a direct comparison in Figure 5.8. Part (a) is the same simulated ZO section as already shown in Figure 5.1(b). Figure 5.8(b) shows the simulated ZO section of the optimized CRS stack after three iterations of residual static correction. Both sections again contain the zoomed part of the second reflection event to visualize the wavelet. In comparison with the wavelet in (a), the wavelet in (b) does not have the same amplitude but the shape of the wavelet is recovered. From the first reflection event, some small time shifts (<5 ms) can still be

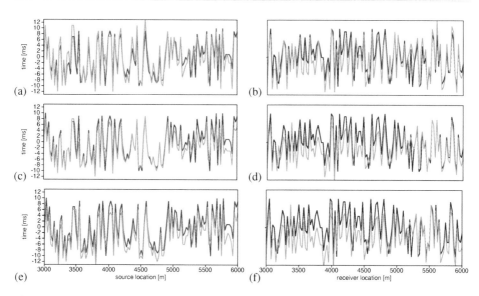

Figure 5.6: Estimated time shifts after iteration 1, 2, and 3. The artificial time shifts added to the multicoverage dataset are shown in red for source locations and in blue for receiver locations. (a), (c), and (e) depict the estimated source static corrections in green for iteration 1, 2, and 3, respectively. (b), (d), and (f) are the estimated receiver static corrections for iteration 1, 2, and 3, respectively.

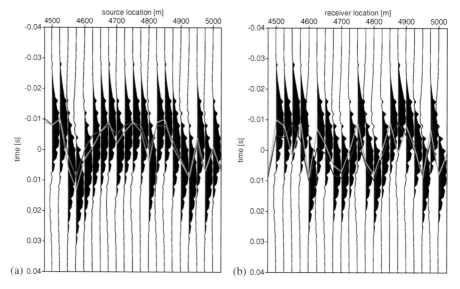

Figure 5.7: Cross correlation stack examples. The cross correlation stacks for a subset of the source locations is displayed in (a) and for a subset of the receiver location in (b). The red and blue curve depict the artificially added source and receiver time shifts, respectively. The green curves represent the estimated time shifts from the cross correlation stacks obtained from the center of the positive lobe around the local maximum closest to a zero time shift.

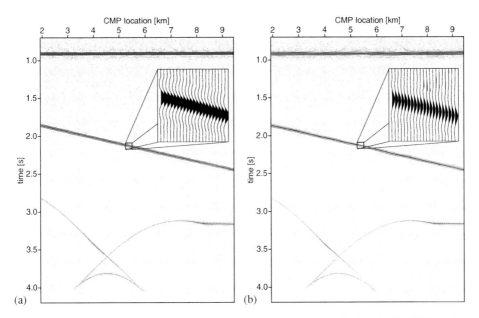

Figure 5.8: Comparison of ZO sections. For comparison, (a) shows again the simulated ZO section without artificial time shifts. (b) displays the simulated ZO section after three iterations of the CRS-based residual static correction method have been applied to the multicoverage dataset distorted by the random but surface-consistent time shifts.

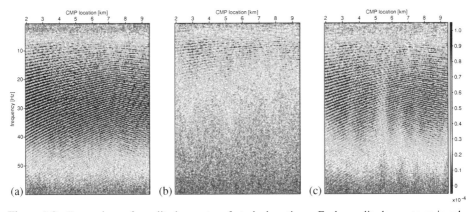

Figure 5.9: Comparison of amplitude spectra of stacked sections. Each amplitude spectrum is calculated from the simulated ZO section obtained as result of the optimized CRS stack but in different stages of the iterative residual static correction and shown in an arbitrary unit. (a) depicts the amplitude spectrum before any artificial time shifts have been applied. (b) shows the amplitude spectrum directly after the artificial time shifts have been added. The frequency content is dramatically reduced compared to (a) due to the artificial time shifts. (c) is the amplitude spectrum after 3 iterations of the CRS-based residual static correction method. In comparison to (a), nearly the whole amplitude spectrum has been recovered with the estimated residual static corrections.

observed. These remaining small undulations are caused by the artificial static time shifts which have been selected from random numbers but are not constrained to have a mean of zero. Thus, the best fitting CRS operator provided by an automatic coherency analysis does not have to provide the correct ZO traveltime for the reflection events at each considered CMP location. This effect is mainly visible for small ZO traveltimes, where the time shifts have a stronger influence in relation to the traveltime itself. Furthermore, the systematic misfit introduced by the second and third iteration have finally decreased the expected improvements.

Nevertheless, the improvements after the residual static corrections are also recognizable in the frequency domain. Therefore, I calculated the Fourier transform of the simulated ZO sections at three stages of the applied processing flow. Figure 5.9(a) depicts the amplitude spectrum of the simulated ZO section provided by the optimized CRS stack applied to the original multicoverage dataset with noise but without the artificial time shifts. As a band limited wavelet has been used during the generation of the dataset and a frequency filter has been applied to obtain the colored noise, the amplitude spectrum mainly contains frequencies between 10 Hz and 50 Hz. The dominant frequency of the wavelet is around 30 Hz. Figure 5.9(b) shows the decrease in the frequency content of the simulated ZO sections obtained directly after the artificial time shifts have been applied to the multicoverage dataset. The amount of this decrease is not constant over the CMP locations. It depends on the extent of the CRS aperture and the time shifts inside this CRS aperture. Finally, Figure 5.9(c) displays the recovered frequency content of the simulated ZO section after three iterations of the CRS-based residual static correction method have been applied to the distorted multicoverage dataset. It is obvious that the amplitude spectrum is not perfectly recovered but this is again caused by the misfit of the estimated static corrections and the influence of this misfit depending on the selected CRS aperture.

Finally, I again want to emphasize that this dataset was intended for a first test, only. Nevertheless, it has already shown the potential of the method. In this case, the obtained estimates from the first iteration quite well fit the artificial time shifts and further iterations can be performed to refine the first estimates.

5.2 Synthetic data example B

Now, the next step towards a more realistic test of the CRS-based residual static correction method is to make use of a real dataset with low complexity. As this dataset is small and requires only little CPU time, I will present here different residual static correction parameter sets applied to this multicoverage dataset while the parameters for the optimized CRS stack remain unchanged during all iterations.

5.2.1 Survey and CRS stack parameters

The used real dataset is a subset of a multicoverage dataset recorded in Northern Germany. This subset consists of 100 CMP gathers where the CMP fold varies between 29 and 31 traces. The multicoverage dataset has an offset range between 60 ft and 3600 ft and a total number of 3028 traces. The provided dataset already underwent routine processing which also included the application of some iterations of a conventional residual static correction method. Within the subset discussed in the following, the source locations vary between 4200 ft and 10800 ft with an average source point spacing of around 120 ft. The receiver locations range between 4200 ft and 10740 ft with a receiver spacing of 60 ft. From the used split-spread acquisition geometry (i. e., receivers with positive and

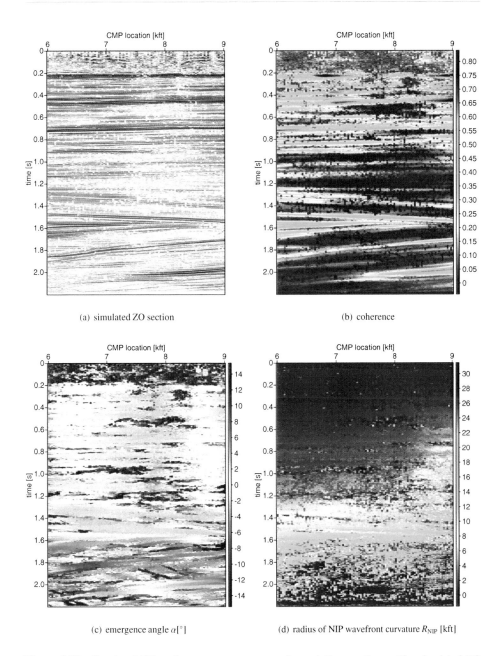

(a) simulated ZO section

(b) coherence

(c) emergence angle $\alpha[°]$

(d) radius of NIP wavefront curvature R_{NIP} [kft]

Figure 5.10: Simulated ZO, coherence, emergence angle, and R_{NIP} sections. The simulated ZO section obtained from the CRS stack directly applied to the provided dataset is shown in (a). (b), (c), and (d) depict the coherence, emergence angle α, and the radius of the NIP wavefront curvature R_{NIP} corresponding to the ZO section in (a).

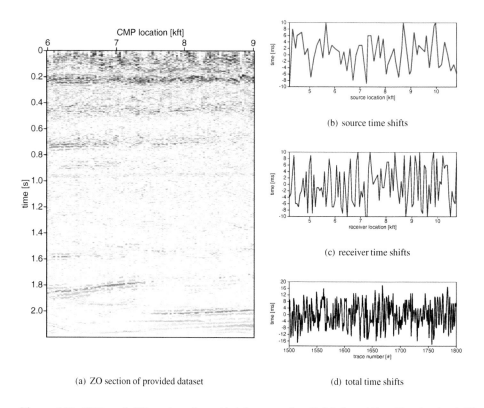

(a) ZO section of provided dataset

(b) source time shifts

(c) receiver time shifts

(d) total time shifts

Figure 5.11: CRS stack ZO section of provided dataset and artificial time shifts. The simulated ZO section of the optimized CRS stack applied to the provided multicoverage dataset is shown in (a). (b) and (c) depict the entire set of the artificial source and receiver time shifts that have been added to the multicoverage dataset. (d) shows a subset of the total time shifts.

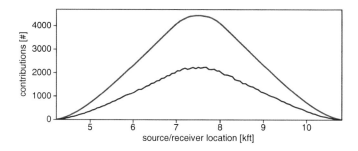

Figure 5.12: Number of contributions to correlation stacks. The red line shows the number of contributions for each source correlation stack and the blue line shows it for each receiver correlation stack.

Context	Processing parameter	Setting
General parameter	Dominant frequency	40 Hz
	Coherence measure	Semblance
	Data used for coherence analysis	Normalized traces
	Temporal width of coherence band	22 ms
Velocity and constraints	Near surface velocity	4000 ft/s
	Tested stacking velocities	2000 ... 20000 ft/s
Target zone	Simulated ZO traveltimes	0 ... 2.2 s
	Simulated temporal sampling interval	2 ms
	Number of simulated ZO traces	101
	Spacing of simulated ZO traces	30 ft
Aperture and taper	Minimum ZO aperture	200 ft @ 0.15 s
	Maximum ZO aperture	3500 ft @ 1.75 s
	Minimum midpoint aperture	400 ft @ 0.15 s
	Maximum midpoint aperture	3700 ft @ 1.75 s
	Relative taper size	30 %
Automatic CMP stack	Initial moveout increment for largest offset	2 ms
	Number of refinement iterations	3
Linear ZO stack	Tested emergence angles	-60 ... 60°
	Initial emergence angle increment	1°
	Number of refinement iterations	3
Hyperbolic ZO stack	Initial moveout increment for largest ZO distance	2 ms
	Number of refinement iterations	3
Hyperbolic CS / CR stack	Initial moveout increment for largest offset	2 ms
	Number of refinement iterations	3
Local optimization	Coherence threshold for smallest traveltime	0.01
	Coherence threshold for largest traveltime	0.005
	Maximum number of iterations	100
	Maximum relative deviation to stop	10^{-4}
	Initial variation of emergence angles	6°
	Initial variation of R_{NIP}	5%
	Initial variation of transformed R_N	6°
	Transformation radius of R_N	100 m

Table 5.3: Synthetic data example B: processing parameters used for the ZO simulation by means of the CRS stack. Some of the listed parameters are not relevant for the residual static correction, but are required for the reproducibility of the CRS processing. For further details, refer to Mann (2002).

---

negative offsets along the 2D seismic line), 56 different source locations and 110 different receiver locations are obtained within the provided multicoverage dataset.

The parameter set used for the optimized CRS stack is listed in Table 5.3. The resulting simulated ZO section directly applied to the provided dataset is shown in Figure 5.10(a). The simulated ZO section is dominated by more or less horizontal plane reflection events. Only few slightly dipping reflection events are clearly visible. The reflections events already show a good reflection event continuity except

Context	Processing parameter	Setting
Input	Data used for moveout correction	Normalized traces
	Moveout correction performed by	Optimized CRS attributes
	ZO section used as pilot traces	Fresnel stack
Cross correlation	Maximum correlation shift	100 ms
	ZO traveltime used for correlation	0.2 … 2.1 s
	Correlation weight	CRS semblance
	Minimum number of 'live' samples per traces	0
Estimation of static correction	Data used for static correction estimation	Original correlation results
	Method applied	Center of positive area around local maximum closest to zero time shift
	Minimum threshold for method applied	30%
	Minimum number of contributions	sources: 0, receiver: 0

Table 5.4: Synthetic data example B: processing parameters used for the CRS-based residual static correction.

for some small gaps which decrease in size with increasing traveltime. These gaps are a bit easier to observe in the coherence section which is displayed in Figure 5.10(b). This behavior can be related to the number of traces contained inside the CRS aperture. For small traveltimes, only few traces are inside the CRS aperture and enter into the coherency-based determination of the CRS attributes. Thus, the coherence analysis might fail due to the low number of traces. Figures 5.10(c) and 5.10(d) display the ZO section of the CRS attributes emergence angle α and the radius of NIP wavefront curvature R_{NIP}, respectively. The angle section shows in compliance with the stacked section in part (a) mainly emergence angles of zero or between +4° and -8°.

In the same way as in the synthetic data example A above, artificial random but surface-consistent time shifts have been applied to the provided multicoverage dataset. These artificial time shifts are again used to simulate a LVL on top of the investigated area. The source and receiver time shifts can again vary between -10 ms and +10 ms. Thus, their sum, i. e., the total time shifts, are randomly distributed between -20 ms and +20 ms. In contrast to the first data example, the dominant frequency has increased from 30 Hz to 40 Hz which yields that the applied time shifts are larger in relation to the dominant wavelength. This makes the correction for the added time shifts a bit more difficult. The simulated ZO section obtained after the artificial time shifts have been added to the pre-stack traces (Figure 5.11(a)) is strongly deteriorated by the artificial static time shifts in such way that much that hardly any reflection event up to 1.8 s can be recognized. I have to emphasize that all ZO sections are displayed with the same amplitude range for an easier comparison against each other. Figures 5.11(b) and 5.11(c) display the entire set of artificial source and receiver time shifts, whereas Figure 5.11(d) shows a subset of the 3028 artificial total time shifts.

5.2.2 Residual static correction

Now, the next step in the testing scheme is to apply the CRS-based residual static correction method to the artificially deteriorated multicoverage dataset. Table 5.4 reveals the basic parameter set which will be changed parameter by parameter, of course, only one parameter at the same time. The following parameters are tested for their influence on the estimation of the residual static corrections after three iterations:

1. the influence of the CRS attributes: before (i.e., initial but smoothed) or after the local optimization (i.e., optimized)

2. the method how to estimate the residual static corrections: the global maximum or the center of the positive lobe around the local maximum closest to zero time shifts

3. the pilot trace: normal stack or Fresnel stack

4. weighting the traces before the cross correlations are calculated: CRS coherence or unweighted

5. normalization of each cross correlation before correlations are stacked: by the cross signal power or as is, i.e., no normalization

6. input for the cross correlations: normalized traces or original traces

Figure 5.13 shows source and receiver static correction estimates, all after the first iteration but with different parameter sets for the application of the CRS-based residual static correction method. To refine the obtained estimates, two more iterations have been performed and the results after the third iteration are displayed in Figure 5.14. This was intended to see whether the results get close to the artificial time shifts and how fast the values converged to the result after the third iteration.

The first two figures (Figures 5.13(a) and 5.13(b)) depict the source and receiver time shifts obtained with the basic parameter set as listed in Table 5.4. These two figures are taken as reference for the comparison with the results obtained with the different parameter sets in the order as mentioned above. Nevertheless, there are still some misfits of the estimated time shifts (green curves) to the artificially added ones (red and blue curve). This can be explained by the provided multicoverage dataset itself. There might still be some residual statics contained in the dataset. Thus, the CRS-based residual static correction method is also applied to the dataset without artificial time shifts later in this chapter. Which parameter set presented in the following yields the best result will also be investigated later on with the comparison of the coherence sections. So far, the estimated static corrections are already very close to the artificial ones except for the locations towards both ends of the subset, where only few traces contributed to the cross correlation stacks. In this case, the lowest number of contributions to the cross correlation stacks at the outermost source or receiver locations was 4. The result obtained with the basic parameter set after three iterations shown in Figures 5.14(a) and 5.14(b) converged very fast as the second and third iteration yielded only few additional time shifts, all smaller than ± 3 ms and ± 1 ms, respectively.

The next parameter set was used with the initial CRS attributes and stacked sections instead of those after the local optimization of the CRS attributes. The difference of Figures 5.13(c) and 5.13(d) after the first iteration compared to the results of the basic parameter set is very small except at both ends of the seismic line. In contrast, the results after the third iteration (Figures 5.14(c) and 5.14(d)) show a larger difference. Furthermore, the convergence of the estimated values is not as fast as obtained from the basic parameter set with optimized CRS attributes. As a conclusion of this comparison, if a first test on a large real dataset has to be conducted to see the influence of residual static corrections, the local optimization can be omitted as it requires much more CPU-time than the initial CRS stack.

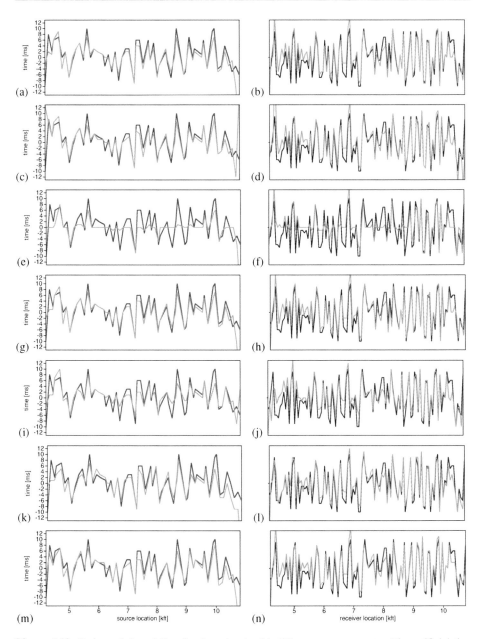

Figure 5.13: Estimated time shifts after iteration 1 with different parameter sets. The artificial time shifts added to the multicoverage dataset are shown in red for source locations and in blue for receiver locations. (a), (c), (e), (g), (i), (k), and (m) depict the estimated source static corrections in green after the first iteration with the basic parameter set and the other parameter sets in the order as listed in the text, respectively. (b), (d), (f), (h), (j), (l), and (n) depict the same but for the receiver locations.

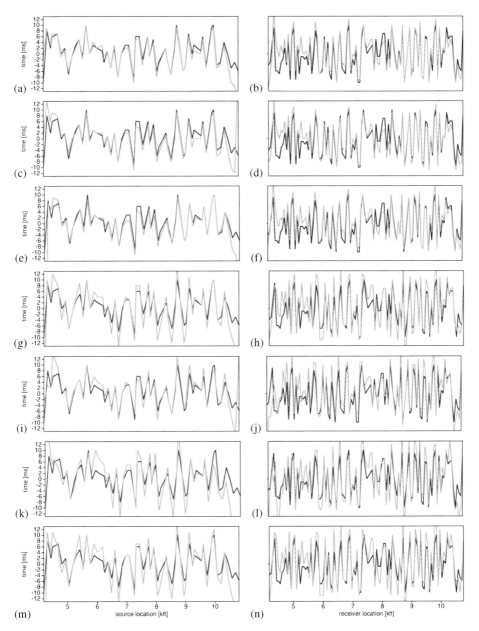

Figure 5.14: Estimated time shifts after iteration 3 with different parameter sets. The artificial time shifts added to the multicoverage dataset are shown in red for source locations and in blue for receiver locations. (a), (c), (e), (g), (i), (k), and (m) depict the estimated source static corrections in green after 3 iterations with the basic parameter set and the other parameter sets in the order as listed in the text, respectively. (b), (d), (f), (h), (j), (l), and (n) depict the same but for the receiver locations.

The initial CRS attributes have been smoothed in an event-consistent way to remove outliers and fluctuations from the CRS attributes before they enter into the local optimization or the CRS-based residual static correction estimation.

Now, the method of extracting the estimated residual static corrections from the cross correlation stacks was changed to the global maximum extraction. From Figures 5.14(e) and 5.14(f), the influence of the extraction method becomes not as clear as from Figures 5.13(e) and 5.13(f). After the first iteration, there are many source and receiver locations with an estimated time shift of zero. This is caused by the asymmetry of the positive lobes, where the global maximum is mainly located at zero time shifts. Here, the separation of source and receiver static corrections did not work as well as for some other locations. Nevertheless, the non-zero time shifts at some locations pushed the results for the subsequent iterations in the right direction. Thus, the results after the third iteration are again close to the artificial ones. But the convergence of the values is even worse than for using the initial sections. Consequently, more iterations are required to end up with a stable final result for the estimated residual static corrections. Furthermore, it is not guaranteed that the time shift of some source or receiver locations help to find the most accurate solution of the residual static problem as there is, in general, no unique solution. However, the synthetic data example A has shown that the extraction of the global maximum in the second and third iteration should not result in a systematic shift of the traces. Thus, it is recommended that at least for the first iteration, the center of the positive lobes should be used. Afterwards, the cross correlation stacks can help to decide which extraction method should be used in further iterations.

For parameter set 3, the parameter concerning the pilot traces has been changed to take the ZO traces of the so-called normal stack as pilot traces. In the basic parameter set, the pilot traces are taken from the Fresnel stacked section, i. e., the CRS aperture has been shrunk to the projected Fresnel zone. This influences the stacked amplitudes in such a way that if the stack operator leaves the positive lobes of the wavelet as it only approximates the true moveout dependency of reflection events, the negative lobes or noise are summed up and decrease the stacked amplitude. This will also influence the cross correlations which depends on the amplitudes as also indicated by Figure 4.9. Figures 5.13(g) and 5.13(h) show only small differences compared to the results obtained with the basic parameter set. The estimated residual static corrections fit very well to the artificially added ones. After the third iteration (Figures 5.14(g) and 5.14(h)), the results show a larger misfit compared to the artificial time shifts. I assume that this misfit is mainly influenced by the pilot traces as the Fresnel stack shows a bit higher frequency content compared to the normal stack. This directly influences the results of each cross correlation and also the stacked cross correlations. The maxima for the Fresnel stacked traces as pilot traces are shorter. Therefore, the estimated static corrections can be determined in a more accurate way.

The next parameter set was changed in the way to weight the traces before the cross correlations are performed. The basic parameter set used the CRS coherence, whereas for this parameter set the traces (within the CRS aperture) are unweighted. As the CRS coherence can be regarded as a quality measure for the reliability of reflection events, it can be used to increase the influence of reflection events during the calculation of the cross correlations. However, the coherence varies from reflection event to reflection event. Thus, in the worst case, i. e., the CRS coherence is dominant along one reflection event, it can influence the cross correlations that much that a cross correlation containing only this dominating reflection event would yield the same results. Nevertheless, Figures 5.13(i) and 5.13(j) depict that the estimated residual static corrections after the first iteration are not as close to the artificial ones as for the basic parameter set. As Figure 5.11(a) shows that only some reflection events are recognizable with a low S/N ratio after the artificial time shifts have been added, the CRS coherence becomes very important. The more iterations are performed and the smaller the estimated time shifts after each iteration, the less important the coherence will be. The resulting estimated time shifts after the third iteration (Figures 5.14(i) and 5.14(j)) are a bit different to the estimates from the basic parameter set but still close to the artificial ones. Thus, it is recommended to use the CRS coherence as an additional weight for the subsequent cross correlations.

The cross correlation results can additionally be normalized by the power of the correlated traces. This normalization equalizes the influence of traces inside each CRS super gather before the cross correlation stacks are performed. After the first iteration, Figures 5.13(k) and 5.13(l) show a better fit to the artificial time shifts than the results of the basic parameter set at many source and receiver locations. After the third iteration (Figures 5.14(k) and 5.14(l)), the misfit gets larger. As already mentioned, the provided real dataset might still contain some remaining residual statics to which my artificial time shifts have been added. Such a normalization can introduce artefacts in the cross correlation stacks. This can happen in cases where the time window used for the cross correlation is only partly contained within the considered CRS aperture. As the CRS aperture increases in midpoint and offset direction with larger traveltime, the more traces are inside this aperture. Limiting these traces with the time window used for the cross correlations, on the one hand, can reduce the number of traces that contribute to the cross correlation stacks and on the other hand, can reduce the number of so-called live sample the closer the traces are to the borders of the CRS aperture. In the worst case, such a trace can contain only one remaining live sample. Thus, the cross correlation of the pilot trace with this sample can yield (depending on the signs of the correlated amplitudes) a maximum at the location of the largest reflection event amplitude at the corresponding time shift. If the normalization is applied to such cross correlation results, this normalization can overweight such results and produce artefacts in the cross correlation stacks. As long as these artefacts are located at larger time shifts than the positive lobe which will yield the desired time shift, these artefacts are meaningless when the local maximum closest to zero time shift or the center of the positive lobe around this local maximum are considered. But if the artefacts are contained inside the positive lobe, the estimated residual static correction can be shifted to a wrong value. To further avoid such artefacts due to the normalization, a minimum number of live samples contained in the correlated traces is introduced and can be defined by the user.

The last test with one changed parameter has been performed where the normalization before the traces enter into the cross correlations was turned off. Thus, the original traces as provided in the multicoverage dataset are used for the cross correlations. The normalization can be regarded as a kind of trace balancing. Thus, the amplitude changes along reflection events will decrease. This trace balancing has the effect of a weight that also reduces the variations considering the cross correlation results. Figures 5.13(m) and 5.13(n) show the residual static corrections after the first iteration which match pretty well the artificially added time shifts. Compared with the basic parameter set result after the first iteration, the results of parameter set 6 vary a bit more around the artificial values. Looking at the results after the third iteration (Figures 5.14(m) and 5.14(n)) shows even larger variations around the artificial time shifts than the results of the basic parameter set.

In Figures 5.15 and 5.16, the undistorted ZO section obtained from the CRS stack applied to the provided dataset is compared with the ZO sections obtained from the CRS-based residual static correction method with the basic parameter set and the parameter sets as listed above applied to the artificially distorted dataset. Each ZO sections displays the result after the application of three iterations with a new search for the attributes in each iteration. All presented ZO sections are the so-called Fresnel stacks from the optimized CRS stack, except for Figure 5.15(c) which shows the Fresnel stack of the initial CRS stack as the local optimization has been omitted and except for Figure 5.16(a), where the

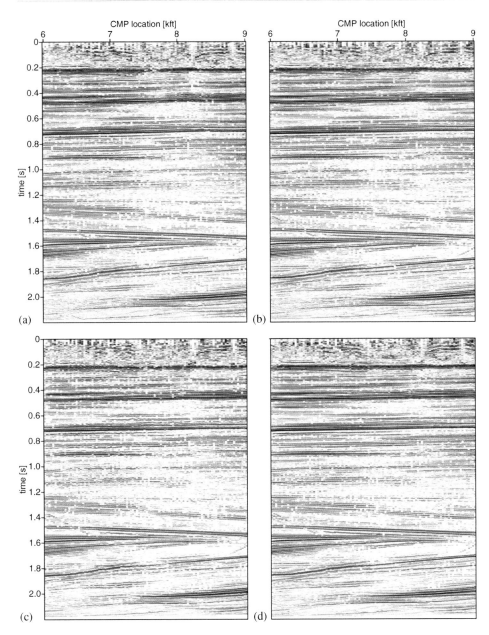

Figure 5.15: Comparison of ZO sections with different parameter sets (part 1). (a) shows again the simulated ZO section obtained from the optimized CRS stack applied directly to the provided dataset. (b) depicts the ZO section after three iterations with the basic parameter set. (c) displays the resulting ZO section after three iterations with parameter set 1. Here, the initial CRS stack is shown. (d) depicts the ZO section obtained after three iterations with parameter set 2.

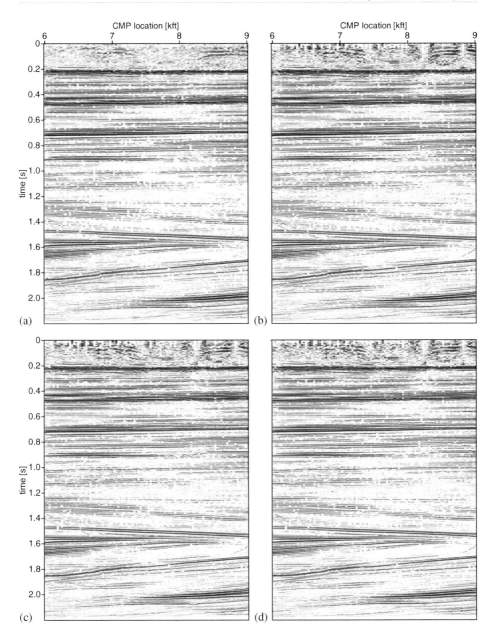

Figure 5.16: Comparison of ZO sections with different parameter sets (part 2). (a) displays the ZO section of the normal stack after three iterations with parameter set 3. (b) depicts the resulting ZO section after three iterations with parameter set 4. (c) shows the ZO section after three iterations with parameter set 5. Last but not least, (d) displays the ZO section obtained after three iterations with parameter set 6.

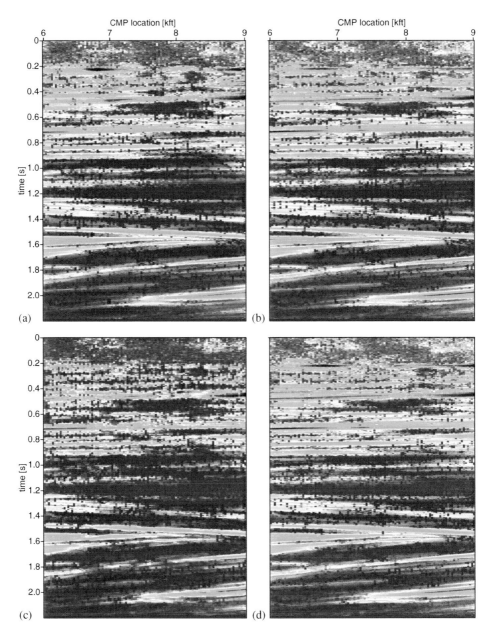

Figure 5.17: Comparison of coherence sections with different parameter sets (part 1). The corresponding CRS coherence sections are shown in the same order as for Figures 5.15. The colormaps are the same for all presented coherence sections and correspond to the colormap of Figure 5.10(b).

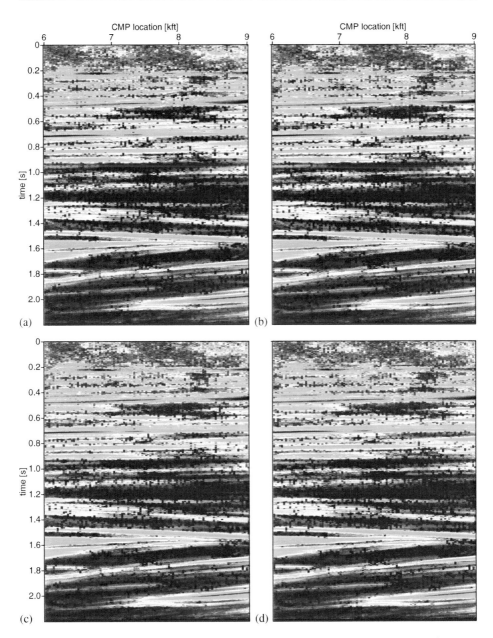

Figure 5.18: Comparison of coherence sections with different parameter sets (part 2). The corresponding CRS coherence sections are shown in the same order as for Figures 5.16. The colormaps are the same for all presented coherence sections and correspond to the colormap of Figure 5.10(b).

95

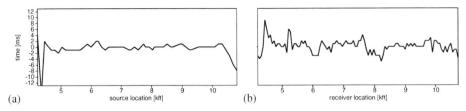

(a) source location [kft]   (b) receiver location [kft]

Figure 5.19: Estimated time shifts after 3 iterations applied to the provided dataset without artificial time shifts. The estimated source static corrections are shown in (a) and the estimated receiver corrections in (b).

normal stack served as pilot traces instead of the Fresnel stacked traces. Thus, Figure 5.16(a) shows the normal stack of the optimized CRS stack.

As the differences are not always visible within the stacked sections, I present also the CRS coherence sections in Figures 5.17 and 5.18 in the same order as the simulated ZO sections are shown. Together with the coherence sections, it becomes obvious that using the initial CRS attributes for the CRS-based residual static correction yields the worst result in these comparisons. Nevertheless, the CRS-based method was still able to correct for the artificially added time shifts and to almost fully recover the multicoverage dataset as it was provided. With the basic parameter set, the stacked section (see Figure 5.15(b)) has been nearly perfectly recovered to match the stacked section directly obtained from the provided dataset (see Figure 5.15(a)). Looking at small traveltimes, the coherence section (Figure 5.17(b)) already indicates that there could be some remaining residual statics in the provided dataset as the coherence along the visible reflection events is less varying than for the original provided dataset. Even with the parameter set 2, where the global maxima of the cross correlation stacks have been used to extract the estimated residual static corrections, the stacked section (Figure 5.15(d)) shows an improved S/N ratio and an enhanced reflection event continuity which can also be observed by the CRS coherence (Figure 5.17(d)). The last four parameter sets show only slight differences in their CRS coherence section (see Figures 5.18(a) - 5.18(d)). Here, the main differences can be easier observed in the stacked ZO section shown in Figures 5.16(a) - 5.16(d). The reflection events between the more prominent ones are easier to distinguish from the noise in the sections of Figures 5.16(a) and 5.16(b) than in the sections of Figures 5.16(c) and 5.16(d). Thus, it is hard to decide which parameter setting should be preferred. It is also possible that changing only one parameter for the presented tests deteriorates the results for the first iteration which the other parameters then compensated for in the subsequent second and third iteration. It is the decision of the interpreter which parameters should be used for further iterations. This decision should be based on the intermediate results after each iteration.

The provided dataset has underwent several preprocessing steps and also a conventional residual static correction as mentioned above. Nevertheless, the comparisons of Figures 5.15 - 5.18 have shown that this dataset still might contain some remaining residual static corrections. This dataset has been also discussed in the work of Kirchheimer (1990), where a CMP-based residual static correction method has been applied to this dataset perturbed by synthetic receiver static time shifts. Irrespectively of the applied preprocessing, the CRS-based residual static correction method is now applied to the provided dataset with the basic parameter set as listed in Table 5.4. After three iterations, the estimated residual static corrections are presented in Figures 5.19. The estimated source static corrections (Fig-

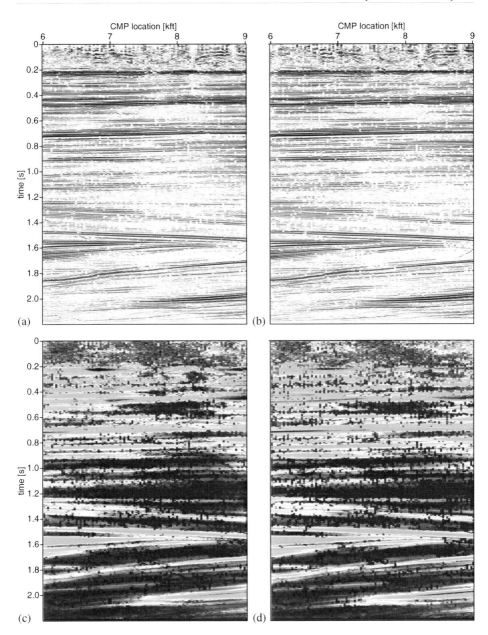

Figure 5.20: ZO and coherence sections before and after three iterations applied to the provided dataset. (a) shows again the simulated ZO section of the optimized CRS stack method applied directly to the provided dataset, whereas (b) displays the stacked section after three iterations of the CRS-based residual static correction method applied to the provided dataset without any artificial time shifts added. (c) and (d) depict the corresponding CRS coherence sections to (a) and (b), respectively.

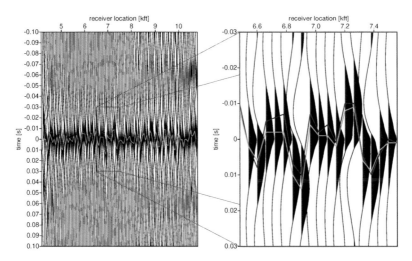

Figure 5.21: Cross correlation stack for receiver locations. This cross correlation has been obtained after the first iteration with the basic parameter set applied to the artificially distorted dataset. The right part shows a zoomed part. The cross correlation stack results vary smoothly but with asymmetrical positive lobes.

ure 5.19(a)) are mainly varying between -1 ms and +2 ms except for the locations close to the ends of the seismic line due to the low coverage, whereas the estimated receiver static corrections (Figure 5.19(b)) are varying between -4 ms and +4 ms.

As conclusion from these tests with different parameter sets, the parameters have to be chosen carefully depending on the considered dataset. There are some parameters that could show their advantages from the second and further iterations like the method to extract the estimates for the residual static corrections in case of rough cross correlation maxima. In this case, the cross correlation stacks vary smoothly but with asymmetrical positive lobes as shown in the right part of Figure 5.21. Other parameter sets are well suited for a first test which is used to obtain results in a short processing time. This can be achieved by omitting the local multi-parameter optimization of the CRS stack. For the real dataset presented above, the optimization ranged between 15 minutes and 35 minutes CPU time[1] while the initial stack ranged between 6 minutes and 8 minutes CPU time. To save processing time by the performed cross correlations, the maximum shift can be reduced. This can be achieved after the first iteration by taking the cross correlation stacks into account (see Figure 5.21: in this case, a maximum time shift of 40 ms would be sufficient). The CPU time needed for the estimation of the residual static corrections ranged between 2 minutes and 3 minutes for each iteration. All CPU-times mentioned here are not representative due to small number of total traces within the multicoverage dataset. For the final processing, I recommend to rely more on the optimized CRS attributes than on the initial CRS attributes. The optimized CRS attributes are usually more accurate than the initial ones due to the local optimization. Thus, as final remark for the synthetic tests presented in this chapter, I recommend to use a parameter set similar to the basic parameter set as listed in Table 5.4 for at least the first iteration of the application of the CRS-based residual static correction method.

[1]CPU time measured on a Pentium M with 1.7 GHz.

Chapter 6

Real data examples

After the synthetic examples in the previous chapter, the CRS-based residual static correction approach is now applied to real datasets. Real data example A contains two recorded seismic lines parallel to each other, whereas real data example B contains only one recorded seismic line but with a larger extension compared to real data example A. Furthermore, real data example A serves as input for an expanded test on real data as a contractor has provided the results of a conventional residual static correction method applied to both lines of this dataset. Unfortunately, conventional results for real data example B are not available. Nevertheless, this second real dataset was acquired on a rough top-surface topography and is used in the following to discuss the application of the CRS-based residual static correction method where the top-surface topography is considered within the CRS stack method itself.

6.1 Real data example A

As mentioned above, this onshore real multicoverage dataset contains two separately recorded seismic lines which have been acquired along two parallel lines in the Upper Rhine Graben in Southern Germany with a distance of about 2 km and a length of 12 km. In the area around both lines, the Rhine Graben shows elevation variations of up to several meters, only, which in relation to the extension of the lines can be neglected for the further processing. The receivers have been layed out statically along both seismic lines with a receiver spacing of 50 m, i. e., the receivers did not move with the sources. The sources (here: three vibrators) moved from one end to the other end of both seismic lines to obtain the two multicoverage datasets. The source spacing has also been 50 m. With this acquisition geometry, the offsets vary between -10 km and +12 km, whereas the CRS stack makes use of offsets up to 2 km, only. In the following, one dataset is called line A and the other line B. The preprocessing (e. g., deconvolution, trace balancing, etc.) has been performed by a contractor for an energy company active in the Rhine Graben. From this company, I kindly got the permission to present some results of the CRS-based imaging workflow here. In addition to the preprocessed multicoverage dataset suitable for the CRS-based imaging workflow, also the residual static corrections of a conventional residual static correction method were provided by the contractor. In the following, I will discuss the results I obtained after three iterations of the CRS-based residual static correction method, whereby the CRS attributes have been searched for in each iteration. The CRS-based residual static correction method is only one part of the entire CRS-based imaging workflow. Figure 6.1 shows a comparison of a conventional reflection seismic imaging workflow and the CRS-based imaging workflow. Hereby, I want

Context	Processing parameter	Setting
General parameter	Dominant frequency	30 Hz
	Coherence measure	Semblance
	Data used for coherence analysis	Normalized traces
	Temporal width of coherence band	30 ms
Velocity and constraints	Near surface velocity	1700 m/s
	Tested stacking velocities	1500 ... 4000 m/s
Target zone	Simulated ZO traveltimes	0 ... 3 s
	Simulated temporal sampling interval	2 ms
	Number of simulated ZO traces	Line A: 413 and line B:428
	Spacing of simulated ZO traces	25 m
Aperture and taper	Minimum ZO aperture	150 m @ 0.2 s
	Maximum ZO aperture	500 m @ 1.2 s
	Minimum midpoint aperture	200 m @ 0.2 s
	Maximum midpoint aperture	2000 m @ 1.2 s
	Relative taper size	30 %
Automatic CMP stack	Initial moveout increment for largest offset	2 ms
	Number of refinement iterations	3
Linear ZO stack	Tested emergence angles	-60 ... 60°
	Initial emergence angle increment	0.5°
	Number of refinement iterations	3
Hyperbolic ZO stack	Initial moveout increment for largest ZO distance	2 ms
	Number of refinement iterations	3
Hyperbolic CS / CR stack	Initial moveout increment for largest offset	2 ms
	Number of refinement iterations	3
Local optimization	Coherence threshold for smallest traveltime	0.001
	Coherence threshold for largest traveltime	0.0005
	Maximum number of iterations	100
	Maximum relative deviation to stop	10^{-4}
	Initial variation of emergence angles	6°
	Initial variation of R_{NIP}	5%
	Initial variation of transformed R_N	6°
	Transformation radius of R_N	100 m

Table 6.1: Real data example A: processing parameters used for the ZO simulation by means of the CRS stack. Some of the listed parameters are not relevant for the residual static correction, but are required for the reproducibility of the CRS processing. For further details, refer to Mann (2002).

to emphasize that only some steps of both workflows in the time domain are discussed in this thesis. For further descriptions of the CRS-based imaging workflow, please refer to Hertweck et al. (2003), Mann et al. (2003), and Heilmann et al. (2004).

The first step for the CRS-based residual static correction method which coincides with the first step of the CRS-based imaging workflow is to apply the CRS stack to the provided multicoverage dataset. In this first CRS stack, a search for the CRS attributes is performed and this provides a simulated ZO

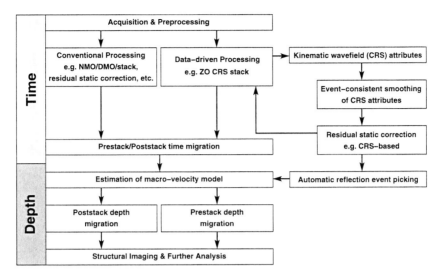

Figure 6.1: Reflection seismic imaging workflow. Displayed are only some steps of a conventional imaging workflow (left hand side) and, in more details, some steps of the CRS-based imaging workflow. The arrow from the CRS-based residual static correction back to the data-driven processing emphasizes the iterative application of residual static correction methods.

---

section as preliminary result. In this case, the optimized CRS stack has been applied to determine the most reliable CRS attributes available for the subsequent CRS-based residual static correction method from the provided dataset. The simulated ZO section of the optimized CRS stack and the corresponding CRS coherence section are shown for line A in Figures 6.4(a) and 6.4(b) and for line B in Figures 6.10(a) and 6.10(b), respectively. The parameters used for the search for the CRS attributes up to the local multi-parameter optimization of the CRS stack processing are listed in Table 6.1. These parameters remained unchanged during all subsequently performed iterations.

Figure 6.4(a) shows the simulated ZO section for line A with a fairly good S/N ratio for the most prominent reflection events. Nevertheless, in between these reflection events the other reflection events are not so easy to distinguish from the noise. Furthermore, there are mainly two areas, around CMP #50 and around CMP #290, that are assumed to be heavily distorted due to fault-controlled fracturing. Both areas are partly contained in the two subsets denoted with 'A' and 'B' which will be discussed later on. The corresponding CRS coherence of the optimized CRS stack shown in Figure 6.4(b) depicts that most of the visible reflection events have a significantly higher CRS coherence than the noise. As mentioned above, the source and receiver elevations have been neglected which introduces some small errors. To compensate for these errors and remaining influences of the weathering layer, residual static correction methods have been applied. In case of line A, a conventional method and the CRS-based method have been applied separately and, additionally, in a combination, i. e., first the conventional method followed by the CRS-based method. For line B, only the separate application of both methods will be discussed here. From the simulated ZO section for line B shown in Figure 6.10(a) and its CRS coherence in Figure 6.10(b), the same effect as for line A can be observed. In contrast to line A, line B is not as strongly affected by fault-controlled fracturing. Thus,

Context	Processing parameter	Setting
Input	Data used for moveout correction	Original traces
	Moveout correction performed by	Optimized CRS attributes
	ZO section used as pilot traces	Fresnel stack
Cross correlation	Maximum correlation shift	40 ms
	ZO traveltime used for correlation	0.8 ... 4.1 s
	Correlation weight	CRS semblance
	Minimum number of 'live' samples per traces	0
Estimation of static correction	Data used for static correction estimation	Original correlation results
	Method applied	Center of positive area around local maximum closest to zero time shift
	Minimum threshold for method applied	30%
	Minimum number of contributions	sources: 0, receiver: 0

Table 6.2: Real data example A: processing parameters used for the CRS-based residual static correction.

only one subset denoted 'C' will be discussed later on.

6.1.1 Residual static correction

As the top-surface topography has been neglected in the processing, a residual static correction should be applied to compensate at least for some errors due to small undulations of the source/receiver locations and also due to the effect of the weathering layer. The residual static corrections for both seismic lines are compared in Figures 6.2 for the source locations and in Figures 6.3 for the receiver locations. Hereby, the provided conventional results for the source and receiver locations are displayed as black curves in all figures. Figure 6.2(a) directly compares the source time shifts estimated after three iterations of the CRS-based residual static correction method applied to the uncorrected dataset (gray curve) with the conventional result in case of line A. The parameters used for all iterations of the CRS-based method are listed in Table 6.2. In some parts, both curves almost perfectly match. However, there are areas where only the shape of the estimated time shift curves remains the same. This can be an effect of the inherent ambiguity of the residual static problem. The same behavior can be observed for the receiver time shifts after three iterations of the CRS-based method compared to the conventional result as shown in Figure 6.3(a). This leads to the assumption that both residual static correction methods provided solutions of the residual static problem that are more or less the same except for some locations. There, the different methods to estimate the time shifts might consider the data differently weighted which is an implementational aspect but can also depend on the used parameters. Thus, it cannot be expected that both curves match perfectly. Nevertheless, another test has been performed. As the conventional results are available, it is possible to correct the multicoverage dataset with the conventionally determined time shifts and afterwards apply the CRS-based method in addition. This test has been performed only for line A. The results of this combined approach are displayed in Figure 6.2(b) for the source locations and in Figure 6.3(b) for the receiver locations. Again, three iterations of the CRS-based method have been applied, but now to the conventionally corrected multicoverage dataset. Comparing the resulting source and receiver time shifts with the conventional time shifts shows that the final estimates (gray curve), i.e., results after conventional plus CRS-based method have been applied, deviate from the conventional result mainly at those locations where the results of separately applied methods are different, too. This also implies that the CRS-based residual static corrections method tends to a different solution in such areas. To decide which solution yields the best results is the responsibility of the interpreter. Nevertheless, the CRS coherence section can help to evaluate the quality of the obtained results irrespectively which residual static correction method has been used.

Figures 6.2(c) and 6.3(c) depict the estimated source and receiver time shifts of the conventional method as black curve and after three iterations of the CRS-based method as gray curve. Both methods have been applied separately to the uncorrected dataset of line B. Except for a few source and receiver locations, the difference of the conventional and the CRS-based result is smaller than 2 ms. As already mentioned for line A, in the areas of larger misfits between the two curves the overall shape still remains the same.

So far, only the estimated time shifts have been discussed assuming that at least one of the presented curves should represent the most convenient result to solve the residual static problem. In the following, I consider the simulated ZO section and their corresponding CRS coherence sections obtained from the different applications of residual static correction methods. Hereby, I firstly discuss the results for line A. I just want to emphasize that also the conventionally corrected dataset has been processed using the CRS stack to obtain the simulated ZO section of the optimized CRS stack suitable for comparison with the other sections. Figure 6.4(a) displays the simulated ZO section before any residual static correction has been applied to the provided dataset, whereas Figure 6.4(b) shows the corresponding CRS coherence section. For comparison, each simulated ZO section and each CRS coherence section is displayed with the same clip and the same colormap. Comparing the simulated ZO section after the dataset has been corrected with the conventional time shifts (see Figure 6.5(a)) shows an improved S/N ratio along the most prominent reflection events. Furthermore, the other reflection events in between the most prominent are easier to distinguish from the noise. It also seems that the faults around CMP #200 are now easier to trace from top to bottom. Considering the CRS coherence in Figure 6.5(b), the conventional time shifts increased almost all coherence values associated with reliable reflection events.

In contrast to the conventional results, the CRS-based residual static correction has been applied to the uncorrected multicoverage dataset. The simulated ZO section shown in Figure 6.6(a) displays almost the same improvement as observed by the conventional results. This can be expected as the estimated time shifts are almost the same, too. Nevertheless, the mainly small differences also result in differences in the stacked sections. This is not as obvious in the stacked section itself but in the CRS coherence section. Figure 6.6(b) depicts the corresponding CRS coherence after three iterations. At first sight, the CRS coherence looks like the CRS coherence after the conventional correction of residual static correction. But taking a closer look at some reflection events reveals that the CRS-based residual static corrections have a higher coherence than after the conventional corrections. The last pair of figures concerning line A (Figures 6.7(a) and 6.7(b)) show the simulated ZO section and its CRS coherence after three iterations of the CRS-based method have been applied to the multicoverage dataset already corrected with the conventional time shifts. There are also only small differences visible compared with the result obtained from the separate application of both methods. Thus, the subsets A and B are used to investigate the differences in more details.

In subset A, I investigate the reflection event from CMP #1 to CMP #100 at traveltimes around 1.6 s. Figures 6.8 depict all simulated ZO section subsets and their CRS coherence sections side by side. It

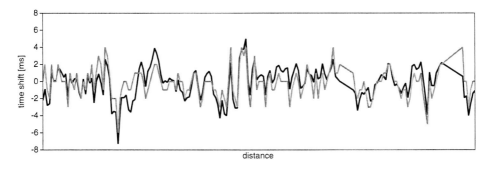

(a) Line A: source static corrections (conventional or CRS-based)

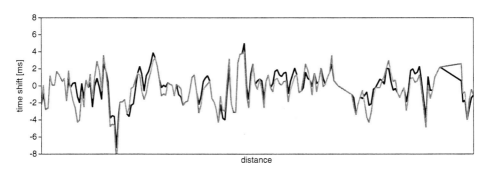

(b) Line A: source static corrections (conventional plus CRS-based)

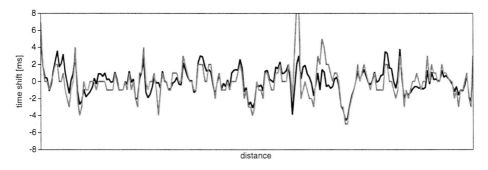

(c) Line B: source static corrections (conventional or CRS-based)

Figure 6.2: Source residual static corrections. The conventional residual static corrections are shown in black. (a) shows the comparison of the conventional (black) and three iterations of the CRS-based method (gray), both applied separately to the uncorrected dataset. (b) depicts the results after three iterations of the CRS-based method applied to the conventionally corrected dataset. (a) and (b) are performed on the dataset from line A, whereas (c) displays the comparison of the conventional (black) and three iterations of the CRS-based method (gray), both applied separately to the uncorrected dataset of line B.

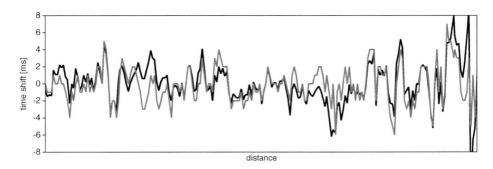

(a) Line A: receiver static corrections (conventional or CRS-based)

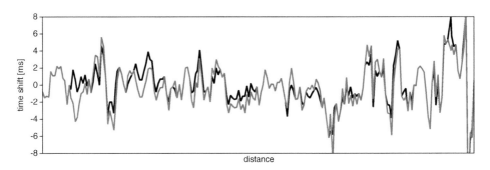

(b) Line A: receiver static corrections (conventional plus CRS-based)

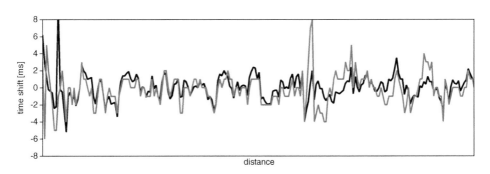

(c) Line B: receiver static corrections (conventional or CRS-based)

Figure 6.3: Receiver residual static corrections. The conventional residual static corrections are shown in black. (a) shows the comparison of the conventional (black) and three iterations of the CRS-based method (gray), both applied separately to the uncorrected dataset. (b) depicts the results after three iterations of the CRS-based method applied to the conventionally corrected dataset. (a) and (b) are performed on the dataset from line A, whereas (c) displays the comparison of the conventional (black) and three iterations of the CRS-based method (gray), both applied separately to the uncorrected dataset of line B.

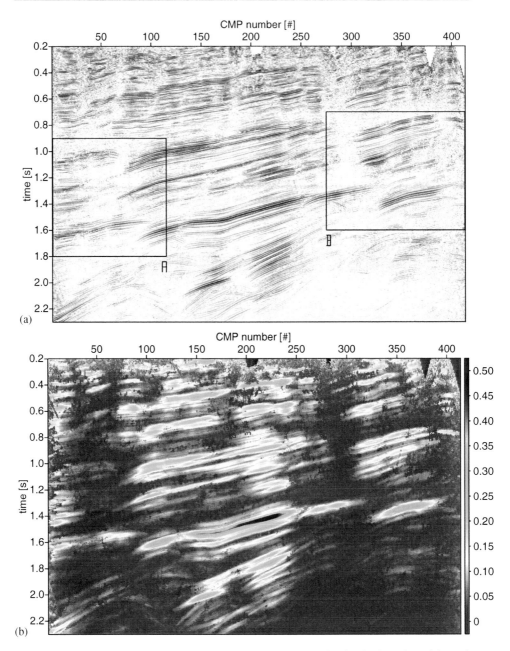

Figure 6.4: Line A before residual static correction. (a) depicts the simulated ZO section of the optimized CRS stack before residual static corrections have been applied. (b) displays the corresponding CRS coherence section. The zoomed sections A and B are displayed in Figure 6.8 and 6.9.

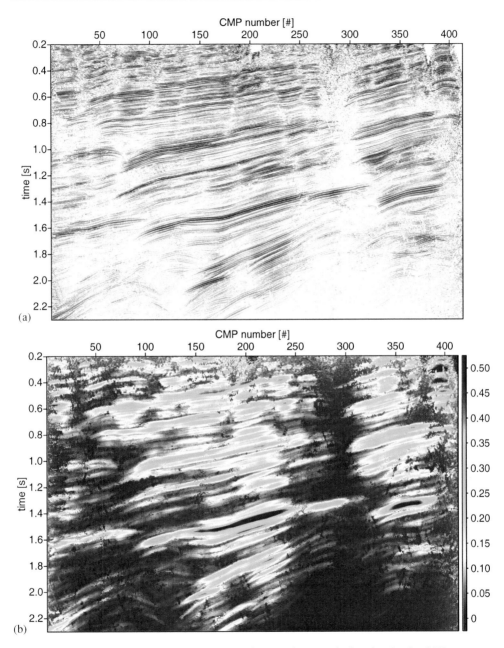

Figure 6.5: Line A after conventional residual static correction. (a) depicts the simulated ZO section obtained from the optimized CRS stack applied to the dataset after conventional residual static correction has been used. (b) displays the corresponding CRS coherence section.

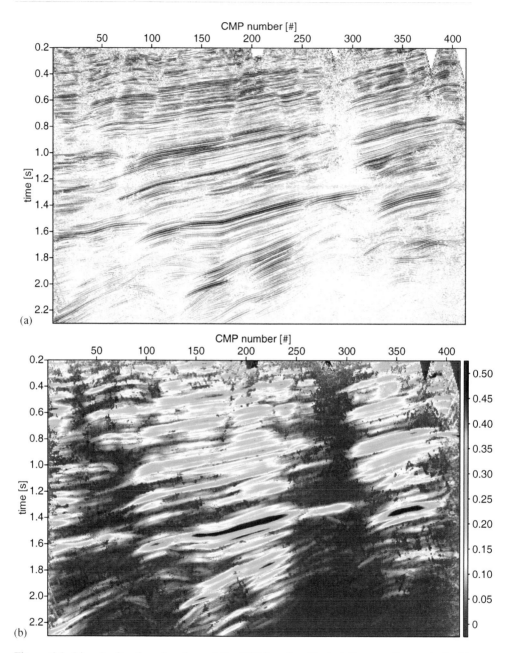

Figure 6.6: Line A after three iterations of the CRS-based residual static correction method. (a) depicts the simulated ZO section of the optimized CRS stack after three iterations have been performed on the uncorrected dataset with the CRS-based residual static correction method. (b) displays the corresponding CRS coherence section.

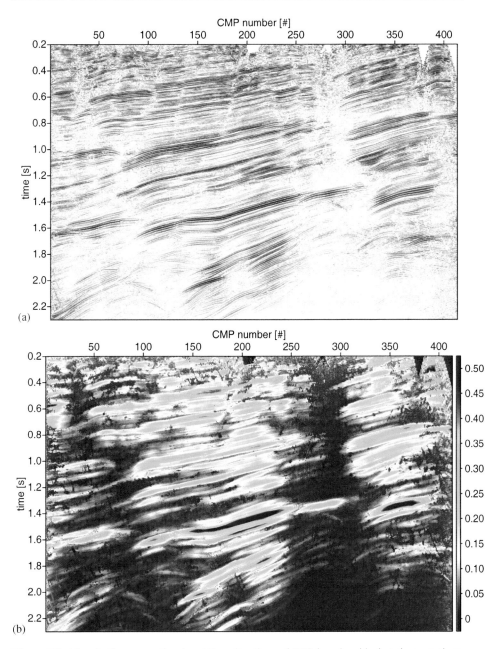

Figure 6.7: Line A after conventional and three iterations of CRS-based residual static corrections. (a) depicts the simulated ZO section of the optimized CRS stack after conventional and three additional iterations of the CRS-based residual static corrections have been applied. (b) displays the corresponding CRS coherence section.

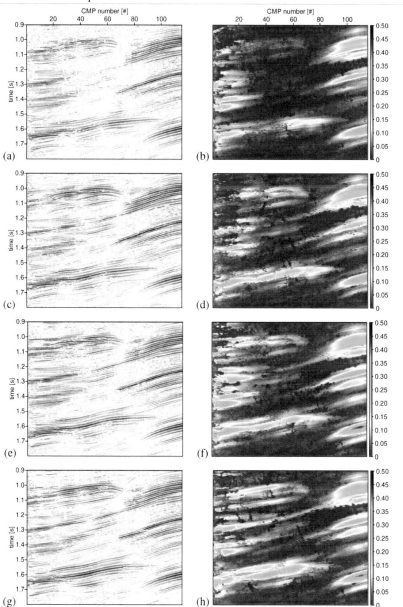

Figure 6.8: Comparison of zoomed area A within line A. (a) shows the subset before residual static corrections have been applied. (c) depicts the result obtained from the optimized CRS stack applied to the dataset after conventional residual static correction. (e) displays the subset after three iterations of the CRS-based method have been performed on the uncorrected dataset. (g) shows the subset after three iterations of the CRS-based method additionally applied to the conventionally corrected dataset. (b), (d), (f), and (h) display the corresponding CRS coherence sections.

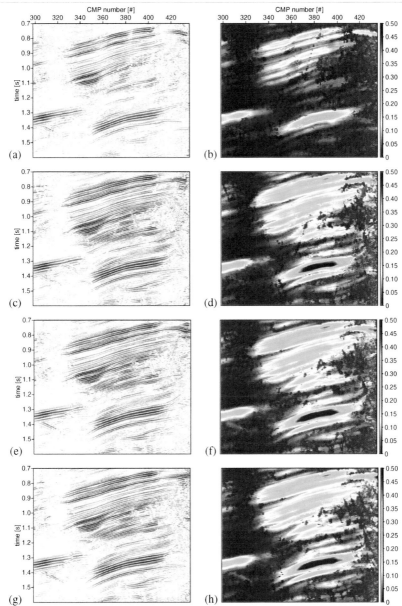

Figure 6.9: Comparison of zoomed area B within line A. (a) shows the subset before residual static corrections have been applied. (c) depicts the result obtained from the optimized CRS stack applied to the dataset after conventional residual static correction. (e) displays the subset after three iterations of the CRS-based method have been performed on the uncorrected dataset. (g) shows the subset after three iterations of the CRS-based method additionally applied to the conventionally corrected dataset. (b), (d), (f), and (h) display the corresponding CRS coherence sections.

becomes obvious that the residual static corrections are necessary to improve the S/N ratio and the reliability of the reflection events as indicated by the CRS coherence. However, the combination of both methods yields the best result in this comparison for the mentioned reflection event. Nevertheless, some of the other reflection events are better visible after only one residual static correction method has been applied.

The subset B shown in Figure 6.9 was chosen with the aim to take a closer look at a reflection event with an opposite dip compared to the most other surrounding reflection events. This reflection event is located between CMP #300 and CMP #320 at traveltimes around 1.4 s. In the uncorrected stacked section and the conventionally corrected stacked section, it is hardly visible. The CRS-based residual static correction method was able to enhance the S/N ratio of this reflection event. Also the combination of both methods enhanced the S/N ratio and the CRS coherence of this reflection event, but not to the same extent as the CRS-based method.

The other areas of both subsets from line A show only slight differences, even in the zoomed view. Nevertheless, the comparison of both methods is not always possible due to economic reasons. Furthermore, the differences observed from line A are not significant enough to recommend the application of only one method. The CRS-based method makes use of far more traces to obtain the estimated time shift due to the spatial extent of the CRS aperture which makes the resulting time shifts more reliable. To investigate further advantages and drawbacks of the CRS-based method, the comparison of the results for line B are also presented in the following figures. Hereby, Figures 6.10(a) and 6.10(b) depict the simulated ZO section and the corresponding CRS coherence section obtained from the optimized CRS stack applied to the uncorrected dataset. The subset denoted with 'C' is shown in Figures 6.13 and will be discussed later on.

The results obtained after the conventional time shifts have been used to correct the provided multicoverage dataset for line B are displayed in Figures 6.11(a) and 6.11(b). In comparison with the uncorrected results, it is again obvious that at least one residual static correction method should be applied to the real dataset to improve the S/N ratio and the continuity of reflection events. Additionally taking the results of the CRS-based residual static correction method into account (see Figures 6.12(a) and 6.12(b)) yields that the CRS-based method has enhanced the S/N ratio and the reflection event continuity even more than the conventional method. However, there are also some areas where the CRS-based method did not work as well as the conventional method, e. g., around CMP #150 at traveltimes of 0.5 s or 1.0 s. Taking a closer look at subset C shows that with the CRS-based residual static correction the reflection events can be traced to a larger extent and that the CRS coherence has more improved with the CRS-based method than with the conventional method. This can mainly be observed at the left hand side of subset C. Thus, for line B the CRS-based residual static correction has shown that the obtained static corrections resulted in a significant improvement of the reflection event continuity and the S/N ratio compared to the conventional results. Nevertheless, the results still depend on the chosen parameters and also on the search for the CRS attributes. If the current implementation of the search routine fails to obtain at least some CRS attributes associated with true reflection events, the CRS-based residual static correction will also fail as it strongly depends on the CRS attributes essential for the moveout correction. However, for line B, the CRS-based method provided a more reliable and stable result due to the far more traces contributing to the estimation of the residual static corrections due to the spatial extent of the CRS aperture.

As conclusion from the above presented results, it is the interpreters choice which method will be applied and, moreover, also the economical aspects have to be considered in practice. Nevertheless, the CRS-based method has shown that it yields at least comparable results as the conventional method.

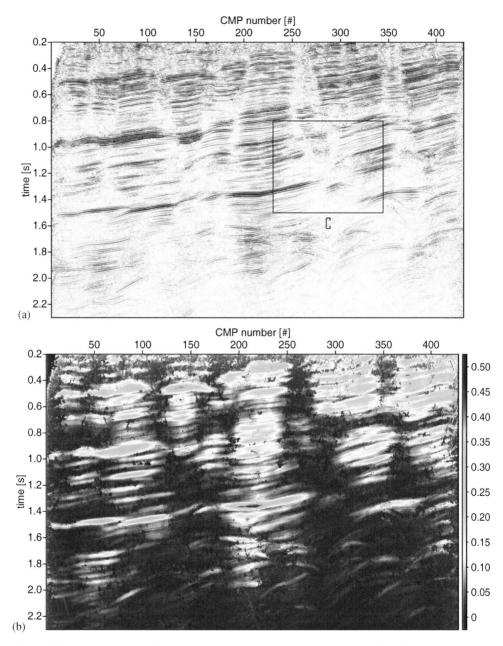

Figure 6.10: Line B before residual static correction. (a) depicts the simulated ZO section of the optimized CRS stack before residual static corrections have been applied. (b) displays the corresponding CRS coherence section.

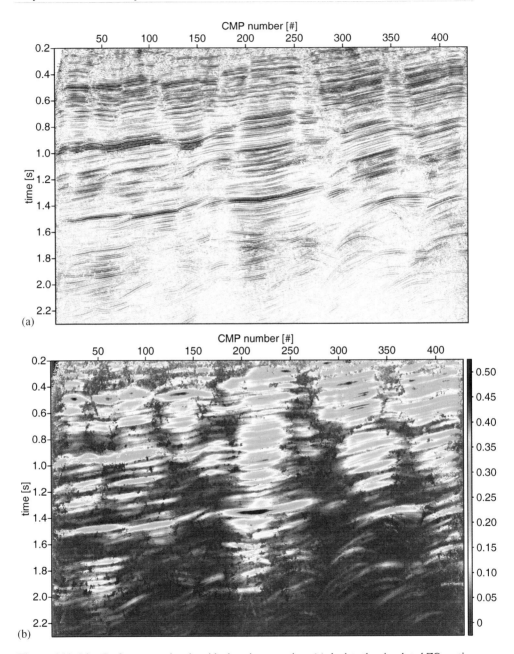

Figure 6.11: Line B after conventional residual static correction. (a) depicts the simulated ZO section of the optimized CRS stack applied to the dataset after conventional residual static corrections have been used. (b) displays the corresponding CRS coherence section.

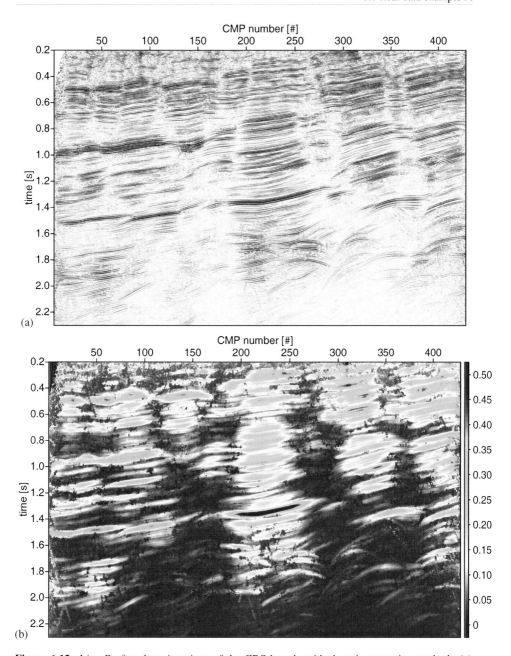

Figure 6.12: Line B after three iterations of the CRS-based residual static correction method. (a) depicts the simulated ZO section of the optimized CRS stack after three iterations have been performed on the uncorrected dataset with the CRS-based residual static correction method. (b) displays the corresponding CRS coherence section.

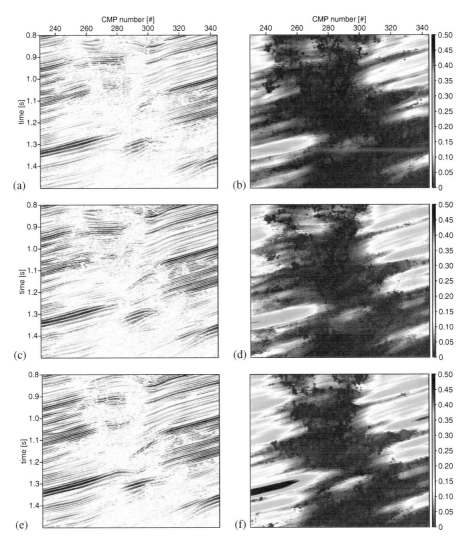

Figure 6.13: Comparison of zoomed area C within line B. (a) shows the subset before residual static corrections have been applied. (c) depicts the result obtained from the optimized CRS stack applied to the dataset after conventional residual static correction. (e) displays the subset after three iterations of the CRS-based residual static correction method have been performed on the uncorrected dataset. (b), (d), and (f) display the corresponding CRS coherence sections.

However, the method makes use of more information obtained from the multicoverage dataset and is a highly automated process, i.e., only little human interaction is necessary. To overcome possible problems with larger residual statics, the use of a priori known/available residual static corrections might be considered in the CRS-based analysis.

6.2 Real data example B

In contrast to all previous examples, the following real dataset has been processed with the CRS stack method considering top-surface topography. The multicoverage dataset has been kindly provided by SAUDI ARAMCO and has been acquired in Saudi Arabia. There, the top-surface topography ranged between 461 m and 595 m above sea level which requires top-surface topography handling. This dataset has already been discussed by von Steht (2004), where the application of the CRS stack method considering top-surface topography and the CRS-based redatuming have been the main targets. With the current implementation, also the application the CRS-based residual static correction method is possible.

The data have been recorded along a seismic line with a length of $\approx$ 40 km. To record each shot along the line, 240 receivers are used with a receiver spacing of 30 m. The receivers moved with the sources in a nearly symmetrical split-spread geometry which yields an offset range of -3602 m to 3607 m. For the total multicoverage dataset, 1279 shots have been carried out with a source spacing of 30 m. Thus, the multicoverage dataset consists of 306960 traces. The traces have been mapped to 2840 CMP bins ranging from CMP #3081 to CMP #5920. The maximum CMP fold is 120. The traces are sampled with 4 ms and the maximum recorded traveltime is 2 s.

The preprocessing (e.g., deconvolution, trace balancing, etc.) has been carried out by a contractor. The provided multicoverage dataset has already been redatumed to the floating datum shown in Figure 6.14 as gray curve. With the preprocessing to the floating datum, a conventional static correction for the weathering layer has been applied which is unfortunately not available for further comparison. The true source and receiver locations are indicated by the black crosses in Figure 6.14. The floating datum has been further smoothed (black curve) for the application of the CRS stack considering the top-surface topography. All elevations are plotted by their projected x location along the regression line through the true x-y locations directly calculated by the 2D ZO CRS stack. The maximum deviation of the true locations to the line of regression is 755 m. In relation to the total length of the seismic line of more than 40 km, the maximum deviation from a straight line is considered as negligible. The differences from the original floating datum to the smoothed floating datum are compensated by field static corrections with a velocity that was chosen constant along the entire line. Note that the used implementation only supports a constant near-surface velocity, but accounting for a varying near-surface velocity is currently in the development stage. The event-consistent smoothing of the CRS attributes has to be applied as a cascaded approach for the 'topographic' CRS stack processing has been used. Here, the CRS stack for smoothly curved measurement surfaces is used to obtain a first estimate of the CRS attributes. Afterwards, these initial CRS attributes served as initial values for the local multi-parameter optimization utilizing the traveltime formula for rough top-surface topography (3.21) (for further details on the cascaded approach refer to von Steht, 2004). The used parameters are listed in Table 6.3 and remained unchanged during the applied CRS-based imaging workflow.

The simulated ZO section obtained from the optimized CRS stack shown in Figure 6.16 is the unre-datumed section belonging to the smoothed floating datum, i.e., it represents the ZO traveltimes

Context	Processing parameter	Setting
General parameter	Dominant frequency	25 Hz
	Coherence measure	Semblance
	Data used for coherence analysis	Normalized traces
	Temporal width of coherence band	84 ms
Velocity and constraints	Near surface velocity	2100 m/s
	Tested stacking velocities	2000 ... 6000 m/s
Target zone	Simulated ZO traveltimes	0.04 ... 2 s
	Simulated temporal sampling interval	4 ms
	Number of simulated ZO traces	2840
	Spacing of simulated ZO traces	15 m
Aperture and taper	Minimum ZO aperture	100 m @ 0.1 s
	Maximum ZO aperture	500 m @ 1.2 s
	Minimum midpoint aperture	150 m @ 0.1 s
	Maximum midpoint aperture	3700 m @ 1.2 s
	Relative taper size	30 %
Automatic CMP stack	Initial moveout increment for largest offset	4 ms
	Number of refinement iterations	5
Linear ZO stack	Tested emergence angles	-60 ... 60°
	Initial emergence angle increment	0.5°
	Number of refinement iterations	5
Hyperbolic ZO stack	Initial moveout increment for largest ZO distance	4 ms
	Number of refinement iterations	3
Hyperbolic CS / CR stack	Initial moveout increment for largest offset	4 ms
	Number of refinement iterations	3
Local optimization	Coherence threshold for smallest traveltime	0.0005
	Coherence threshold for largest traveltime	0.0005
	Maximum number of iterations	100
	Maximum relative deviation to stop	10^{-4}
	Initial variation of emergence angles	6°
	Initial variation of R_{NIP}	5%
	Initial variation of transformed R_N	6°
	Transformation radius of R_N	100 m

Table 6.3: Real data example B: processing parameters used for the ZO simulation by means of the CRS stack. Some of the listed parameters are not relevant for the residual static correction, but are required for the reproducibility of the CRS processing. For further details, refer to Mann (2002).

obtained along central rays emerging at the elevations of the smoothed floating datum. The CRS-based redatuming to a planar reference datum is possible but neither discussed nor presented in the following. For further details concerning the CRS-based redatuming, please refer to Zhang (2003), Heilmann (2002), or von Steht (2004). As the reflection events are almost horizontal in the simulated ZO section, it is evident that they are nearly parallel to the smoothed floating datum for which this section has been simulated. The CRS coherence section corresponding to the simulated ZO section

Context	Processing parameter	Setting
Input	Data used for moveout correction	Normalized traces
	Moveout correction performed by	Optimized CRS attributes
	ZO section used as pilot traces	Fresnel stack
Cross correlation	Maximum correlation shift	50 ms
	ZO traveltime used for correlation	0.3 … 1.8 s
	Correlation weight	CRS semblance
	Minimum number of 'live' samples per traces	0
Estimation of static correction	Data used for static correction estimation	Original correlation results
	Method applied	Center of positive area around local maximum closest to zero time shift
	Minimum threshold for method applied	10%
	Minimum number of contributions	sources: 0, receiver: 0

Table 6.4: Real data example B: processing parameters used for the CRS-based residual static correction.

of Figure 6.16 is shown in Figure 6.18. The CRS coherence values at CMP numbers from 3081 up to 3550 and from 5250 up to 5920 show an acceptable reliability for the observed reflection events. In the area from CMP #3550 up to 4950, the CRS coherence values are quite low but still a correlation with the observed reflection events is possible. However, for the area between CMP #4950 and 5250, the search for the CRS attributes did not detect contiguous and reliable reflection events.

6.2.1 Residual static correction

As mentioned before, field statics are used to redatum the traces to the smoothed floating datum. Thus, the subsequent application of a residual static correction is recommended and the used parameters are listed in Table 6.4. Two runs of estimating the residual static corrections (both runs with the same parameters) have been applied, where the search for the CRS attributes after the first correction has been omitted assuming that the CRS attributes are sufficiently reliable after the first CRS optimization. Nevertheless, after the correction of the pre-stack traces with the estimated time shifts, the pre-stack traces have been stacked again with the unchanged CRS attributes. Furthermore, the CRS coherence has been recalculated with the corrected pre-stack traces. Thus, the input for the second estimation changed in such a way that the pre-stack traces, the pilot traces, and the CRS coherence differ from the first run. As expected, the estimated time shifts after the second run are smaller than in the first run. Only for some locations, there time shifts of around ±5 ms remained, for the most locations the second estimates varied in the range of ±1 ms. After correcting the pre-stack traces with the second estimate of the time shifts, a new search for the CRS attributes has been applied. This procedure was intended to save processing time as the local multi-parameter optimization of the CRS stack method is the most time consuming process. Furthermore, as the second estimate has shown a strong convergence towards zero time shifts, the new search for the attributes can further reduce outliers and might close some gaps along the reflection events.

The estimated time shifts obtained after the two runs of the CRS-based residual static correction method are displayed in Figure 6.15(a) for the source locations and in Figure 6.15(b) for the receiver

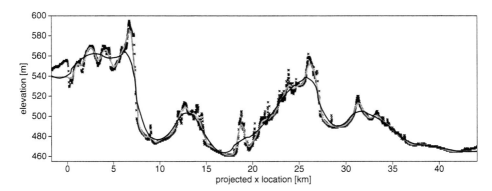

Figure 6.14: Rough and smoothed topography. The black crosses depict the elevations as they have been acquired in the field for all source and receiver locations, whereas the gray curve displays the original floating datum used by a contractor for the preprocessing. The black curve shows the smoothed floating datum that was used for the further processing within the CRS stack.

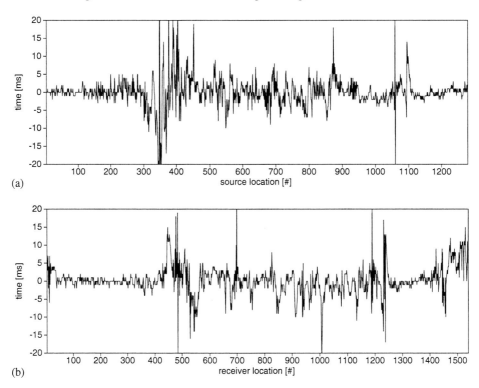

Figure 6.15: Estimated source and receiver static corrections. The shown static corrections have been obtained after two iterations of the CRS-based residual static correction method. (a) shows the source static corrections displayed by the number of the source location and (b) depicts the receiver static corrections also displayed by the number of the corresponding receiver location.

locations. Due to the defined spatial extent of the CRS aperture as listed in Table 6.3, all 1279 source locations and 1540 receiver locations have contributed to the estimation of residual static corrections. The source and receiver time shifts are displayed by the number of the corresponding location. Thus, the change at each location is independent of the distance to its neighboring locations which helps to visualize also small changes. The curves of the estimated time shifts can be correlated with the magnitude of the field static corrections, i.e., the larger the field statics the larger the possible error and, thus, the larger the residual static correction. This can be explained by the constant velocity used for all field static corrections which may not coincide with the local near-surface velocity.

Figure 6.17 shows the simulated ZO section obtained after the local multi-parameter optimization of the CRS stack after the second iteration has been applied, i.e., the two runs of estimating residual static corrections are performed and the resulting corrected multicoverage dataset is used as input for the subsequent new search for the CRS attributes. Comparing this section with the ZO section before the residual static correction shows, at first sight, the increased S/N ratio along almost all observable reflection events. Furthermore, the reflection event continuity has been improved, e.g., around CMP #4050 at traveltime 0.9 s or around CMP #4700 at traveltime 0.5 s. Another effect of the residual static corrections can be mainly observed between CMP #3081 and 3550. There, the undulations along the reflection events have decreased. This effect can be observed the more the smaller the traveltimes. Additionally taking the CRS coherence into account for the comparison underlines the improvements observed in the stacked sections. Here, the CRS coherence increased by a factor of up to 2 along events associated with reliable CRS attributes. In contrast to the other presented examples, I have to mention that the CRS attributes have not been event-consistently smoothed before they entered into the local multi-parameter optimization. The small gaps along the reflection events are assumed to be caused by outliers of the CRS attributes which the optimization could not compensate for. Thus, it is expected that these gaps can be closed by applying the event-consistent smoothing prior to the local multi-parameter optimization.

In addition to the enhanced simulated ZO section, all these changes have a significant impact on the other applications of the CRS-based imaging workflow that make use of, e.g., the CRS attributes like the tomographic approach to build up a smooth macro-velocity model as described by Duveneck (2004). Hereby, the dramatic increase of the CRS coherence along reflection events increases the number of reliable time picks that enter into the calculation of the velocity model. Thus, as a final remark, the CRS-based residual static correction method is a useful tool to further improve or simplify parts of the remaining imaging workflow steps. Furthermore, the highly automated approach reduces the required human interaction. Nevertheless, the parameters have to be chosen carefully depending on the dataset which makes at least some human interaction inevitable.

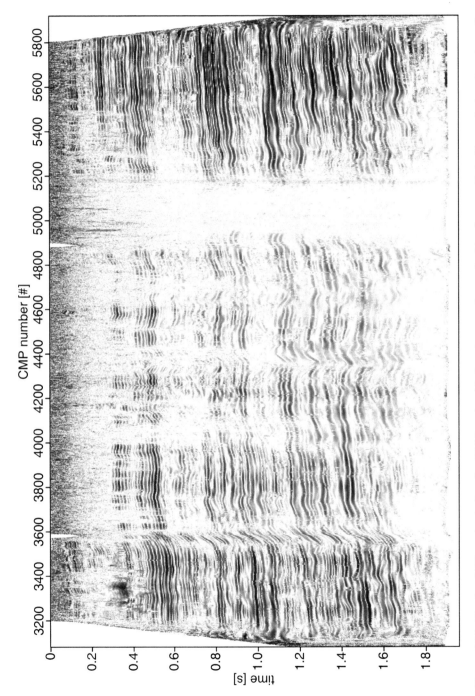

Figure 6.16: Simulated ZO section before the CRS-based residual static correction method has been applied. The stacked section shows the unredatumed result of the optimized CRS stack considering the top-surface topography.

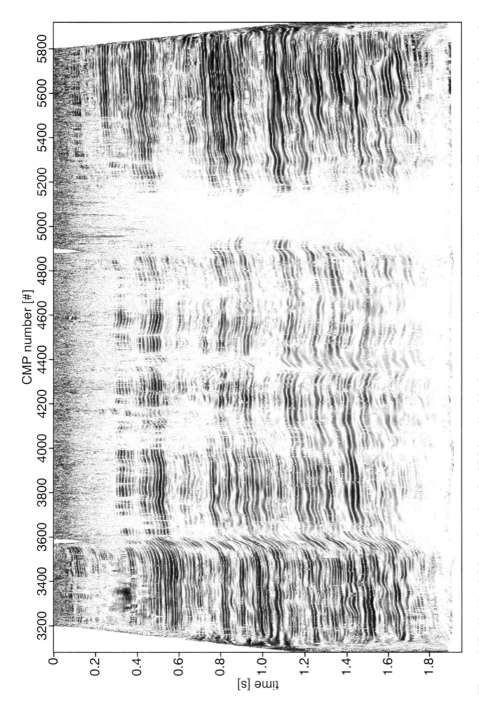

Figure 6.17: Simulated ZO section after the CRS-based residual static correction method has been applied. The stacked section shows the unredatumed result of the optimized CRS stack considering the top-surface topography.

123

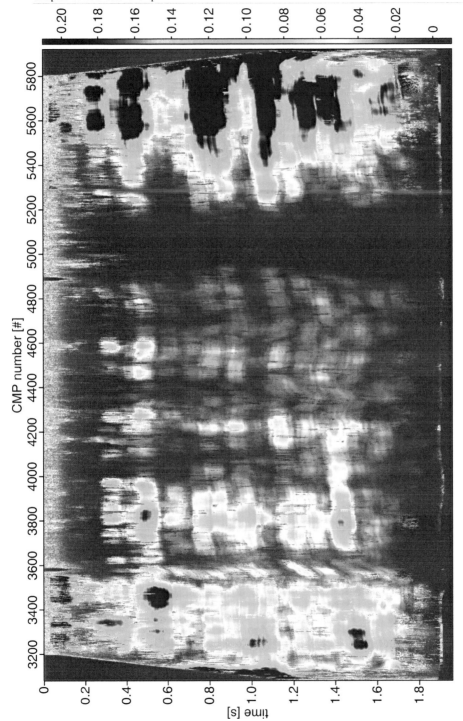

Figure 6.18: CRS coherence section before the CRS-based residual static correction method has been applied. This section is the CRS coherence section of the optimized CRS stack and corresponds to Figure 6.16.

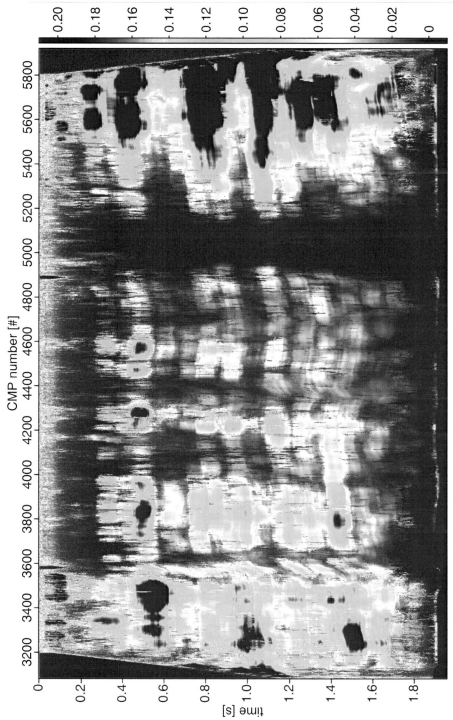

Figure 6.19: CRS coherence section after the CRS-based residual static correction method has been applied. This section is the CRS coherence section of the optimized CRS stack and corresponds to Figure 6.17.

125

Chapter 7

Conclusions and outlook

A small part of the entire reflection seismic imaging workflow has been addressed in the framework of the thesis in hand. This part deals with the estimation of residual static corrections which mainly aims to enhance the S/N ratio and the reflection event continuity in the resulting ZO section and also in the entire pre-stack dataset. Furthermore, residual static corrections are commonly used to eliminate the influence of the weathering layer on the reflection event traveltimes which remains after the datum static correction. For the task of estimating residual static corrections from acquired multicoverage datasets, an entirely data-oriented approach has been extended and utilized. This CRS stack approach directly parameterizes the reflection events in the pre-stack time domain. The parameterization of the common reflection surfaces in depth is performed with the inherent redundancy of the pre-stack data. Thus, the parameters location, orientation, and curvature in depth have not to be explicitly known as the corresponding reflection response in the time domain is described by means of properties of (hypothetical) wavefronts emerging at the acquisition surface.

The theoretical considerations necessary for the parameterization of the wavefronts are based on paraxial ray theory. Furthermore, a geometrical derivation of the used parameterization has been reviewed to gain insight into a geometrical interpretation of the obtained kinematic wavefield attributes, i. e., the CRS attributes. This has been done only for datasets acquired along a seismic line, i. e., the so-called 2D ZO case. As the residual static problem is connected to the weathering layer, the 2D land data case has to be considered. Here, the handling of top-surface topography has been introduced into the CRS stack approach by Zhang (2003) as the measurement surface is, in general, not planar.

The presented extension of the 2D ZO CRS stack approach allows to estimate a solution of the residual static problem. Hereby, the estimation of static time shifts as implemented in the CRS stack method is based on the *stack power maximization method* by Ronen and Claerbout (1985). In contrast to the conventional method, the CRS-based residual static correction method has been adapted for the additional information provided by the CRS stack, i. e., the CRS attributes. As the CRS attributes parameterize the reflection response in the time domain, the CRS attributes can be utilized to eliminate the offset dependency of the parameterized reflection events. This so-called moveout correction is essential for the estimation of the time shifts. Most conventional residual static correction methods rely on an approximate NMO correction which can be regarded as a one-parameter moveout correction, whereas the CRS-based moveout correction is a multi-parameter moveout correction, e. g., three CRS attributes in the 2D ZO case.

The CRS-based method for residual static correction makes use of cross correlations as most conventional methods do. Here, the advantages of the CRS stack are: on the one hand, the higher S/N ratio

127

of the simulated ZO section compared to conventionally obtained ZO sections and, on the other hand, the spatial extent of the CRS aperture. As the traces of the simulated ZO section enter into the cross correlations as pilot traces, also the cross correlation results are expected to improve with the higher S/N ratio compared to, e. g., CMP stacked or NMO/DMO/stacked traces. As already indicated before, the other advantage of the spatial extent of the CRS aperture is that much more traces contribute to the estimation of the residual static corrections. Here, the cross correlation result of each pre-stack trace with the pilot trace is assigned to the corresponding source and receiver locations of the pre-stack traces. Due to the spatial extent of the CRS aperture, several traces inside this aperture share the same source or receiver, respectively. Furthermore, the CRS aperture is used for each CMP location within the target zone which results in an overlap of the CRS apertures depending on their spatial extent. Thus, a cross correlation stack is generated for each source and receiver location with a usually large number of contributions.

The two synthetic data examples proved that the CRS-based residual static correction method is able to estimate time shifts with only few iterations required. The number of iterations strongly depends on the chosen parameter set and requires sometimes experienced human interaction. Furthermore, if conventional residual static corrections have already been applied, the CRS-based residual static correction method can be used to further refine the provided residual static corrections. From the comparison of real data example A, I conclude that the CRS-based method has shown that it is a competitive tool compared to conventional methods. Real data example B has also shown the potential of the presented approach as the CRS coherence has been increased mainly by a factor around two along reliable reflection events. Furthermore, the CRS-based residual static correction method is implemented into the CRS-based imaging workflow which is highly automated.

Besides all these advantages, one still has to keep in mind that these improvements are only possible if the CRS stack method is able to obtain the CRS attributes by its currently implemented search algorithm. Here, I just want to mention that the three CRS attributes are searched for by three one-parameter searches in different subsets of the multicoverage dataset. This cascaded search might fail in some cases which makes a subsequent residual static correction pointless. Furthermore, the CRS stack is based on paraxial ray theory with some assumptions (e. g., high-frequency approximation) which have to be valid for the investigated subsurface structure. Nevertheless, I have shown with several synthetic and real data examples that the extension of the CRS stack method to estimate residual static corrections has the ability to further improve the simulated ZO section and the reliability of reflection events (as indicated by the CRS coherence). Thus, the CRS-based residual static correction method should be considered in the application of the reflection seismic imaging workflow.

Some improvements can still be expected from future developments of the CRS-based residual static correction method. One of such developments mainly deals with the estimation of the residual static corrections. Here, the calculation of the "center of mass" within the positive lobes around the global maximum or the local maximum closest to a zero time shift has been mentioned. With this extraction method for the time shifts, a threshold is not required as in case of some of the already implemented methods and the shape of the positive lobe is accounted for (using equal mass or other mass distributions with the associated time shifts). The presented approach can currently be applied for the 2D ZO case either with planar measurement surface or with rough top-surface topography, only, and has still to be implemented for the 2D finite offset or the 3D case to have the improvements available. There are also some other implementational aspects concerning, e. g., the automation of the iterations which have to be considered in future.

Appendix A

Notation and list of symbols

The chosen variables in the subsequent list are utilized without any physical meaning only to explain the used notation.

$\prod a$ Product symbol $\sum b$ Summation symbol

$$\vec{x} = \begin{pmatrix} x_1 \\ x_2 \\ x_3 \end{pmatrix} \quad \text{Vector}$$

$$\underline{\mathbf{X}} = \begin{pmatrix} x_{11} & x_{12} & x_{13} \\ x_{21} & x_{22} & x_{23} \\ x_{31} & x_{32} & x_{33} \end{pmatrix} = \left(x_{ij} \right) \quad \begin{array}{l} 3 \times 3 \text{ Matrix} \\ (i, j = 1, 2, 3) \end{array}$$

$$\mathbf{X} = \begin{pmatrix} x_{11} & x_{12} \\ x_{21} & x_{22} \end{pmatrix} = \left(x_{ij} \right) \quad \begin{array}{l} 2 \times 2 \text{ Matrix} \\ (i, j = 1, 2) \end{array}$$

$$\underline{\underline{\mathbf{X}}} = \begin{pmatrix} \mathbf{X}_1 & \mathbf{X}_2 \\ \mathbf{X}_3 & \mathbf{X}_4 \end{pmatrix} = \left(x_{ij} \right) \quad \begin{array}{l} 4 \times 4 \text{ Matrix} \\ (i, j = 1, 2, 3, 4) \end{array}$$

$$\vec{x}^T = (x_1, x_2, x_3) \quad \text{Transposed vector}$$

$$\underline{\mathbf{X}}^T = \left(x_{ji} \right) \quad \text{Transposed Matrix}$$

$$\vec{f} = \vec{f}(\vec{x}) = \begin{pmatrix} f_{x_1} \\ f_{x_2} \\ f_{x_3} \end{pmatrix} \quad \text{Vector field}$$

$$f = f(\vec{x}) \quad \text{Scalar field}$$

$$\vec{a} \times \vec{b} \quad \text{Vector product}$$

$$\vec{a} \cdot \vec{b} \quad \text{Scalar product}$$

$$\vec{\nabla} = \begin{pmatrix} \frac{\partial}{\partial x_1} \\ \frac{\partial}{\partial x_2} \\ \frac{\partial}{\partial x_3} \end{pmatrix} \quad \begin{array}{l} \text{Nabla operator} \\ \text{(Cartesian)} \end{array}$$

$$\vec{\nabla}_{\vec{f}} = \begin{pmatrix} \frac{\partial}{\partial f_{x_1}} \\ \frac{\partial}{\partial f_{x_2}} \\ \frac{\partial}{\partial f_{x_3}} \end{pmatrix} \quad \begin{array}{l} \text{Nabla operator} \\ \text{(different variable)} \end{array}$$

$$\frac{df(u(x))}{dx} = \frac{df}{du}\frac{du}{dx} \quad \begin{array}{l} \text{Full derivative} \\ \text{(chain rule)} \end{array}$$

$$\frac{\partial a}{\partial b} \quad \text{Partial derivative}$$

$$\vec{\nabla} \cdot \vec{f} = \frac{\partial f_{x_1}}{\partial x_1} + \frac{\partial f_{x_2}}{\partial x_2} + \frac{\partial f_{x_3}}{\partial x_3} \quad \text{Divergence}$$

$$\vec{\nabla} f = \begin{pmatrix} \frac{\partial f}{\partial x_1} \\ \frac{\partial f}{\partial x_2} \\ \frac{\partial f}{\partial x_3} \end{pmatrix} \quad \text{Gradient}$$

$$\vec{\nabla} \times \vec{f} = \begin{pmatrix} \frac{\partial f_{x_3}}{\partial x_2} - \frac{\partial f_{x_2}}{\partial x_3} \\ \frac{\partial f_{x_1}}{\partial x_3} - \frac{\partial f_{x_3}}{\partial x_1} \\ \frac{\partial f_{x_2}}{\partial x_1} - \frac{\partial f_{x_1}}{\partial x_2} \end{pmatrix} \quad \text{Curl} \qquad \Delta = \left(\vec{\nabla} \cdot \vec{\nabla} \right) = \frac{\partial^2}{\partial x_1^2} + \frac{\partial^2}{\partial x_2^2} + \frac{\partial^2}{\partial x_3^2} \quad \begin{array}{l} \text{Laplace operator} \\ \text{(Cartesian)} \end{array}$$

$$\mathcal{J} = \frac{1}{c} \det \left(\underline{\mathcal{J}} \right) = \frac{1}{c} \left| \frac{d(a_1, a_2, a_3)}{d(b_1, b_2, b_3)} \right| \quad \begin{array}{l} \text{Jacobian} \\ \text{determinant} \end{array} \qquad \underline{\mathcal{J}} = \frac{d(a_1, a_2, a_3)}{d(b_1, b_2, b_3)} = \begin{pmatrix} \frac{da_1}{db_1} & \frac{da_1}{db_2} & \frac{da_1}{db_3} \\ \frac{da_2}{db_1} & \frac{da_2}{db_2} & \frac{da_2}{db_3} \\ \frac{da_3}{db_1} & \frac{da_3}{db_2} & \frac{da_3}{db_3} \end{pmatrix} \quad \text{Jacobian matrix}$$

$$\mathbf{A}^{-1} = \frac{1}{\det(\mathbf{A})} \begin{pmatrix} a_{22} & -a_{12} \\ -a_{21} & a_{11} \end{pmatrix} \quad \text{with} \quad \det(\mathbf{A}) = a_{11}a_{22} - a_{12}a_{21} \quad \begin{array}{l} \text{Inverse matrix} \\ (2 \times 2) \text{ case} \end{array}$$

$$\det(\underline{\mathbf{X}}) = x_{11}x_{22}x_{33} + x_{21}x_{32}x_{13} + x_{31}x_{12}x_{23} - x_{13}x_{22}x_{31} - x_{23}x_{32}x_{11} - x_{33}x_{12}x_{21} \quad \text{Determinant}$$

$$\vec{0} = (0,0,0)^T \quad \text{or} \quad \underline{\mathbf{0}} = \left(0_{ij} \right) \quad \begin{array}{l} \text{Zero vector} \\ \text{or matrix} \end{array} \qquad \underline{I} = \left(\delta_{ij} \right) = \begin{pmatrix} 1 & 0 & 0 \\ 0 & 1 & 0 \\ 0 & 0 & 1 \end{pmatrix} \quad \text{Identity matrix}$$

Important mathematical rule:

- Interchange of differentiation and integration: If $\frac{\partial f(x,t)}{\partial x}$ is continuous the following relation holds:

$$F(x) = \int_a^b f(x,t)\, dt \quad \Longrightarrow \quad \frac{dF(x)}{dx} = \int_a^b \frac{\partial f(x,t)}{\partial x}\, dt$$

A.1   List of abbreviations

The following list contains only the shortcuts used in this thesis and their expanded writings without any explanations:

2D	two-dimensional	FE	finite-element
3D	three-dimensional	LVL	low-velocity layer
CDP	common-depth-point	MZO	migration to zero-offset
CMP	common-midpoint	NIP	normal incidence point
CO	common-offset	NMO/DMO/stack	normal moveout/dip moveout/stack
CR	common-receiver	P-wave	primary wave
CRP	common-reflection-point	RSC	residual static correction
CRS	common-reflection-surface	S-wave	secondary wave
CS	common-source or common-shot	S/N ratio	Signal-to-noise ratio
FD	finite-difference	ZO	Zero-offset

130

Appendix B

Some properties of the ray propagator matrix

The surface-to-surface propagator matrix $\mathbf{T}$ (in 2D) or $\underline{\mathbf{T}}$ (in 3D) and the laws derived from it can be used to obtain approximate solutions for many ray-theoretical problems. One of these problems is the calculation of paraxial traveltimes used within this thesis (see Chapter 2). There, the knowledge of some properties of the propagator matrix comes into play in the treatment of the above mentioned problem which will be reviewed in this appendix.

B.1 Symplecticity

Bortfeld (1989) introduced the surface-to-surface propagator matrix $\underline{\mathbf{T}}$. This matrix is related to the second derivatives of the traveltime. The order of differentiation has to be irrelevant which leads to symmetries of $\underline{\mathbf{T}}$. These symmetries are the so-called symplecticity and this property can be expressed by the inverse matrix of $\underline{\mathbf{T}}$:

$$\underline{\mathbf{T}}^{-1} = \begin{pmatrix} \mathbf{A} & \mathbf{B} \\ \mathbf{C} & \mathbf{D} \end{pmatrix}^{-1} = \begin{pmatrix} \mathbf{D}^T & -\mathbf{B}^T \\ -\mathbf{C}^T & \mathbf{A}^T \end{pmatrix}. \tag{B.1}$$

As $\underline{\mathbf{T}}^{-1}\underline{\mathbf{T}} = \underline{\mathcal{I}}$ has to be fulfilled, the following conditions for $\underline{\mathbf{T}}$ are obtained:

$$\mathbf{B}^{-1}\mathbf{A} = \left(\mathbf{B}^{-1}\mathbf{A}\right)^T, \tag{B.2a}$$

$$\mathbf{D}\mathbf{B}^{-1} = \left(\mathbf{D}\mathbf{B}^{-1}\right)^T, \tag{B.2b}$$

$$\mathbf{A}^T\mathbf{D} - \mathbf{C}^T\mathbf{B} = \mathcal{I}. \tag{B.2c}$$

B.2 Reverse ray

If the start and end points of a paraxial ray are interchanged, $\underline{\mathbf{T}}^*$ is the 4×4 surface-to-surface propagator matrix of the reverse ray. The wavefront still travels along the central ray but now in the opposite direction. To express this fact mathematically, the slowness vectors of the reverse ray have the same components as the corresponding vectors of the original ray except for the opposite sign. The reverse propagator matrix $\underline{\underline{T}}^*$ in terms of the elements of the propagator matrix $\underline{\underline{T}}$ reads (Hubral, 1983; Hubral et al., 1992a)

$$\underline{\underline{T}}^* = \begin{pmatrix} \mathbf{D}^T & \mathbf{B}^T \\ \mathbf{C}^T & \mathbf{A}^T \end{pmatrix} . \tag{B.3}$$

If the central ray is a normal ray, i. e., a ray with normal incidence at the reflection point and the downgoing and upgoing branches are identical, the matrix $\underline{\underline{T}}$ and matrix $\underline{\underline{T}}^*$ also have to be identical which directly leads to the condition

$$\mathbf{A} = \mathbf{D}^T . \tag{B.4}$$

Together with the symplecticity, the number of independent submatrices of matrix $\underline{\underline{T}}$ are reduced to two for a normal central ray. Additionally, it becomes obvious that the submatrices $\mathbf{B}$ and $\mathbf{C}$ are symmetric in such a case.

B.3 Chain rule

For an arbitrary point M along the central ray from point S to point R, the chain rule states that the propagator matrix $\underline{\underline{T}}$ for the entire ray from point S to point R, i. e., $S\,MR$, satisfies the equation

$$\underline{\underline{T}}(S,R) = \underline{\underline{T}}(M,R)\,\underline{\underline{T}}(S,M) . \tag{B.5}$$

$\underline{\underline{T}}(S,M)$ and $\underline{\underline{T}}(M,R)$ denote the propagator matrices for the ray branches from point S to point M and from M to R, respectively. Thus, the entire ray $S\,MR$ is build up by the two branches $S\,M$ and MR. Independent on the choice of M to lie upon an actual reflecting or transmitting interface or an arbitrary ficticious interface, this equation holds for all points M. The propagator matrices of the two branches $\underline{\underline{T}}(S,M)$ and $\underline{\underline{T}}(M,R)$ can be further decomposed with equation (B.5) by replacing the former point M as new start point S or end point R, respectively. Thus, the propagator matrix $\underline{\underline{T}}$ can ultimately be written as a product of many ray-segment propagator matrices. This general decomposition is referred to as chain rule of the propagator matrix $\underline{\underline{T}}$.

Choosing the intermediate point M to be the reflection point of a normal ray and denoting the propagator matrix of the branch from S to M as $\hat{\underline{\underline{T}}}$, the entire propagator matrix $\underline{\underline{T}}$ can be formulated due to equations (B.3) and (B.5) as

$$\underline{\underline{T}} = \hat{\underline{\underline{T}}}^*\hat{\underline{\underline{T}}} . \tag{B.6}$$

The matrix product shows that

$$\underline{\underline{T}} = \begin{pmatrix} \mathbf{A} & \mathbf{B} \\ \mathbf{C} & \mathbf{D} \end{pmatrix} = \begin{pmatrix} \hat{\mathbf{D}}^T & \hat{\mathbf{B}}^T \\ \hat{\mathbf{C}}^T & \hat{\mathbf{A}}^T \end{pmatrix}\begin{pmatrix} \hat{\mathbf{A}} & \hat{\mathbf{B}} \\ \hat{\mathbf{C}} & \hat{\mathbf{D}} \end{pmatrix} = \begin{pmatrix} \hat{\mathbf{D}}^T\hat{\mathbf{A}} + \hat{\mathbf{B}}^T\hat{\mathbf{C}} & \hat{\mathbf{D}}^T\hat{\mathbf{B}} + \hat{\mathbf{B}}^T\hat{\mathbf{D}} \\ \hat{\mathbf{C}}^T\hat{\mathbf{A}} + \hat{\mathbf{A}}^T\hat{\mathbf{C}} & \hat{\mathbf{C}}^T\hat{\mathbf{B}} + \hat{\mathbf{A}}^T\hat{\mathbf{D}} \end{pmatrix} . \tag{B.7}$$

Please note that the last equation restates equation (B.4).

Appendix C

Refraction seismics in relation to the near surface

The boundary between two layers of constant velocities affects the propagation direction of a seismic ray according to Snell's law (see Figure C.1). Here, Snell's law states that the ratio of the sine of the angle with the normal of the interface to the velocity remains constant:

$$\frac{\sin \gamma_I}{v_1} = \frac{\sin \gamma_T}{v_2} = -\frac{\sin \gamma_R}{v_1}.$$ (C.1)

To fulfill Snell's law in the case of increasing velocities with depth, the angle of the refracted ray with respect to the interface normal has to increase, too. Thus, the ray is refracted away from the interface normal. There exists a so-called critical angle γ_C depending on the velocity contrast. For this critical incidence angle $\gamma_I = \gamma_C$, the refracted angle $\gamma_R = 90°$, i.e., $\sin \gamma_R = 1$. The condition for the critical angle follows from equation (C.1) as

$$\sin \gamma_C = \frac{v_1}{v_2}.$$ (C.2)

Consider a wavefront reaching a planar interface under the critical angle: the wavefront is refracted along the interface and propagates with the velocity of the deeper, i.e., faster medium while it emits energy back to the surface. Therefore, at each point along this planar interface, there exists a ray from the interface to the surface with the same incidence angle except for the opposite sign, i.e., $\gamma_R = -\gamma_I$. This can be explained by considering the corresponding wavefront to the refracted ray (see, e.g., Telford et al., 1976) which is known as head-wave or Mintrop wave.

Refraction data is usually represented in the most convenient form by plotting the extracted first-break traveltimes t_x vs. the corresponding source-receiver distance x (see upper part of Figure C.2). In the following, the time-distance relations for a simple case of a horizontal interface separating two constant velocity layers are derived (see lower part of Figure C.2). The horizontal discontinuity is located at depth z_0 and the layer above has the velocity v_1 and the layer below v_2. Now, consider the total time along the refraction path $ABCD$ in the lower part of Figure C.2 which is given by

$$
\begin{aligned}
t_x &= t_{AB} + t_{BC} + t_{CD} = 2t_{AB} + t_{BC} \\
&= 2\frac{z_0}{v_1 \cos \gamma_C} + \frac{x - 2z_0 \tan \gamma_C}{v_2} \\
&= \frac{2z_0}{v_1 \cos \gamma_C} - \frac{2z_0 \sin \gamma_C}{v_2 \cos \gamma_C} + \frac{x}{v_2},
\end{aligned}
$$ (C.3)

133

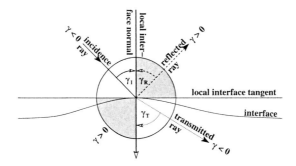

Figure C.1: Snell's law and sign convention. The emerging ray is depicted as solid black arrow and its angle γ_I is measured against the interface normal. The angles of the reflected ray (dashed arrow) γ_R and of the refracted ray (dotted arrow) γ_T are also measured against the interface normal.

---

or, expressed in terms of velocities,

$$t_x = \frac{x}{v_2} + \frac{2z_0 \sqrt{v_2^2 - v_1^2}}{v_1 v_2}. \tag{C.4}$$

This is the mathematical form of a straight line in a time-distance plot with the slope of $\frac{1}{v_2}$ and the intersection with the time axis is at the so-called intercept time

$$t_i = \frac{2z_0 \sqrt{v_2^2 - v_1^2}}{v_1 v_2}. \tag{C.5}$$

The second straight line in the upper part of Figure C.2 is simply the direct arrival of the wavefront propagating along the measurement surface to all receivers. Its slope is $\frac{1}{v_1}$ and its mathematical description is given by

$$t_x = t_{AD} = \frac{x}{v_1}. \tag{C.6}$$

The intersection of both lines is called the crossover distance and reads

$$x_O = 2z_0 \sqrt{\frac{v_2 + v_1}{v_2 - v_1}}. \tag{C.7}$$

At the crossover distance x_O, the refracted and direct waves need the same time to reach this location but along different ray paths. At smaller offsets the direct wave firstly reaches the receivers whereas at larger offsets the refracted wave firstly arrives at the receivers. Thus, it is necessary to have receivers at sufficiently small offsets to observe the direct wave in the data. The velocities of the upper and lower layer can be directly estimated from the first arrivals in the time-distance plot.

Now, the depth z_0 of the interface can be calculated from the intercept time (C.5) which is solved for z_0 and reads

$$z_0 = \frac{t_i v_1 v_2}{2 \sqrt{v_2^2 - v_1^2}}. \tag{C.8}$$

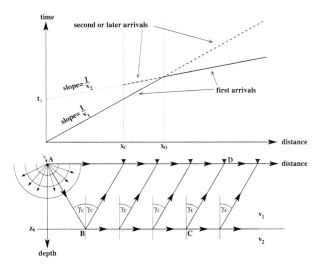

Figure C.2: Refraction seismic scheme. Upper part: time-distance plot with the first arrivals given by the direct arrivals until the crossover distance x_O is reached and afterwards given by the refracted arrivals. Lower part: ray paths for the direct wave and the refracted wave in case of a horizontal plane reflector and increasing velocities with depth, i. e., $v_1 < v_2$.

The critical distance depicted by x_C is the minimal distance from which the refracted wave might occur in the recordings of the receivers. There, the ray emerges at the reflector with the critical angle. The critical distance can be expressed as

$$x_C = \frac{2z_0v_1}{\sqrt{v_2^2 - v_1^2}} \, . \tag{C.9}$$

In real data, it is often difficult to use first breaks to estimate the intercept time and the velocities of the uppermost layers, i. e., the velocity of the first layer v_w which is also called the weathering layer and the velocity of the next layers of the bedrock v_{b_i}. This difficulties can be caused by, e. g., the undulation of the base of the weathering layer or due to a low number of receivers at offsets smaller than the crossover distance x_O. There exist many different methods to overcome these problems which incorporate far more sophisticated descriptions of the subsurface structure. One descriptive method commonly used was introduced by Hagedoorn (1959). It is known as the Plus-Minus method. In practice, it is common to assume a constant velocity for the weathering layer. With this assumption, the variations of the base of weathering depth can be estimated with the Plus-Minus method.

Please note that problems occur if the principle of refraction seismics is explained with zero-order ray theory. The kinematic aspects are not affected, but the dynamics are not fully covered. The energy of the incident wavefront dispenses along refracted wavefronts. Consequently, the energy along a refracted wavefront is very small and could usually not be observed in seismic data. Therefore, other wavefronts have to produce the seismic signal at the same traveltime as the refracted wavefronts. This can be caused by constructive interference. Nevertheless, this appendix was intended to explain only the principle of refraction seismics and how it can provide more information about the weathering layer expressed on the simple example of a horizontal plane reflector with constant velocity layers. For a dipping or even more complicated cases, please refer to Cox (1999).

List of Figures

List of Tables

References

Aki, K. and Richards, P. G. (1980). *QUANTITATIVE SEISMOLOGY Theory and Methods*, volume 1. W. H. Freeman and Company, San Francisco.

Bergler, S. (2001). Common-Reflection-Surface stack for common offset - theory and application. Master's thesis, Karlsruhe University.

Bleistein, N. (1984). *Mathematical Methods for Wave Phenomena*. Academic Press Inc., New York.

Bleistein, N. (1986). Two-and-One-Half Dimensional In-Plane Wave Propagation. *Geophysical Prospecting*, 34:686–703.

Born, M. and Wolf, E. (1959). *Principle of optics*. Pergamon Press Inc.

Bortfeld, R. (1989). Geometrical ray theory: Rays and traveltimes in seismic systems (second-order approximations of the traveltimes). *Geophysics*, 54:342–349.

Buttkus, B. (2000). *Spectral Analysis and Filter Theory in Applied Geophysics*. Springer-Verlag, Berlin.

Červený, V. (2001). *SEISMIC RAY THEORY*. Cambridge University Press, Cambridge.

Cox, M. J. G. (1999). *Static Corrections for Seismic Reflection Surveys*. Society of Exploration Geophysicists, Tulsa, Oklahoma.

Deregowski, S. M. (1986). What is DMO? *First Break*, 4(7):7–24.

Duveneck, E. (2004). *Tomographic determination of seismic velocity models with kinematic wavefield attributes*. Logos Verlag, Berlin.

Ewig, E. (2003). Theory and application of residual static corection by means of CRS attributes. Master's thesis, Karlsruhe University.

Fitch, A. A. (1981). *Developments in geophysical exploration methods*. Applied Science publishers Ltd.

Gajewski, D. (1998). Determining the ray propagator from traveltimes. In *Expanded Abstracts*, pages 1900–1903. 71st Annual International Meeting, Society of Exploration Geophysicists. ST15.

Hagedoorn, J. G. (1959). The plus-minus method of interpreting seismic refraction sections. *Geophysical Prospecting*, 7:158–182.

Hale, D. (1991). *Dip Moveout Processing*. Society of Exploration Geophysicists. Course notes.

Heilmann, Z. (2002). The Common-Reflection-Surface Stack under Consideration of the Acquisition Surface Topography and of the Near-Surface Velocity Gradient. Master's thesis, Karlsruhe University.

Heilmann, Z., Mann, J., Duveneck, E., and Hertweck, T. (2004). CRS-Stack-Based Seismic Reflection Imaging – A Real Data Example. In *Expanded Abstracts*. 66th Annual International Meeting, European Association of Geoscientists & Engineers. Session P211.

Hertweck, T., Jäger, C., Mann, J., and Duveneck, E. (2003). An integrated data-driven approach to seismic reflection imaging. In *Expanded Abstracts*. 65th Annual International Meeting, European Association of Geoscientists & Engineers. Session P004.

Höcht, G. (1998). Common-Reflection-Surface stack. Master's thesis, Karlsruhe University.

Höcht, G., de Bazelaire, E., Majer, P., and Hubral, P. (1999). Seismics and optics: hyperbolae and curvatures. *Journal of Applied Geophysics*, 42(3,4):261–281.

Hubral, P. (1983). Computing true amplitude reflections in a laterally inhomogeneous earth. *Geophysics*, 48(8):1051–1062.

Hubral, P. and Krey, T. (1980). *Interval velocities from seismic reflection traveltime measurements*. Society of Exploration Geophysicists.

Hubral, P., Schleicher, J., and Tygel, M. (1992a). Three-dimensional paraxial ray properties, Part I: Basic relations. *Journal of Seismic Exploration*, 1:265–279.

Hubral, P., Schleicher, J., and Tygel, M. (1992b). Three-dimensional paraxial ray properties, Part II: Applications. *Journal of Seismic Exploration*, 1:347–362.

Hubral, P., Schleicher, J., Tygel, M., and Hanitzsch, C. (1993). Determination of Fresnel zones from traveltime measurements. *Geophysics*, 58(5):703–712.

Jäger, R. (1999). The Common-Reflection-Surface stack - theory and application. Master's thesis, Karlsruhe University.

Kelly, K. R., Ward, R. W., Treitel, S., and Alford, R. M. (1976). Synthetic seismograms: A finite-difference approach. *Geophysics*, 41(1):2–27.

Kirchheimer, F. (1990). Residual statics by CDP-localized stack optimization. *Geophysical Prospecting*, 38:577–606.

Kirkpatrick, S., Gelatt, C. D., J., and Vehhi, M. P. (1983). Optimization by simulated annealing. *Science*, 220:671–680.

Kravtsov, Y. A. and Orlov, Y. I. (1990). *Geometrical Optics of Inhomogeneous Media*. Springer Verlag, New York.

Levin, F. K. (1971). Apparent velocity from dipping interface reflections. *Geophysics*, 36:510–516.

Mann, J. (2002). *Extensions and Applications of the Common-Reflection-Surface Stack Method*. Logos Verlag, Berlin.

Mann, J. and Duveneck, E. (2004). Event-consistent smoothing in generalized high-density velocity analysis. In *Expanded Abstracts*, pages 2176–2179. 74th Annual International Meeting, Society of Exploration Geophysicists. ST1.1.

Mann, J., Duveneck, E., Hertweck, T., and Jäger, C. (2003). A sesismic reflection imaging workflow based on the Common-Reflection-Surface stack. *Journal of Seismic Exploration*, 12:283–295.

Mann, J. and Höcht, G. (2003). Pulse Stretch Effects in the Context of Data-Driven Imaging Methods. In *Extended Abstracts*. 65th Annual International Meeting, European Association of Geoscientists & Engineers. Session P007.

Mann, J., Jäger, R., Müller, T., Höcht, G., and Hubral, P. (1999). Common-Reflection-Surface stack – a real data example. *Journal of Applied Geophysics*, 42:301–318.

Müller, T. (1999). *The Common-Reflection-Surface stack method - seismic imaging without explicit knowledge of the velocity model*. PhD thesis, Karlsruhe University.

Neidell, N. S. and Taner, M. T. (1971). Semblance and other coherency measures for multichannel data. *Geophysics*, 36(3):482–497.

Popov, M. M. (2002). *Ray Theory and Gaussian Beam Method for Geophysicists*. Universidade Federal da Bahia, Salvador.

Popov, M. M. and Pšenčík, I. (1978). Computation of ray amplitudes in inhomogeneous media with curved interfaces. *Studia geophys. et geod.*, 22:248–258.

Profeta, M., Moscoso, J., and Koremblit, M. (1995). Minimum field static corrections. *The Leading Edge*, 14(6):684–687.

Reynolds, J. M. (1997). *An Introduction to Applied and Environmental Geophysics*. John Wiley & sons, Chichester, England.

Ronen, J. and Claerbout, J. F. (1985). Surface-consistent residual statics estimation by stack-power maximization. *Geophysics*, 50(12):2759–2767.

Roth, M. (2004). Seismic processing: past, present and future. *First Break*, 22(8):53–57.

Rothman, D. H. (1985). Nonlinear inversion, statistical mechanics, and residual statics estimation. *Geophysics*, 50(12):2784–2796.

Rothman, D. H. (1986). Automatic estimation of large residual statics corrections. *Geophysics*, 51(2):332–346.

Sheriff, R. E. (2002). *Encyclopedic Dictionary of Applied Geophysics*. Society of Exploration Geophysicists, Tulsa. Fourth Edition.

Stockwell Jr., J. W. (1995). 2.5-D wave equations and high frequency asymptotics. *Geophsics*, 60(1):556–562.

Strang, G. and Fix, G. (1973). *An analysis of the finite element method*. Prentice Hall.

Taner, M. T., Koehler, F., and Alhilali, K. A. (1974). Estimation and correction of near-surface time anomalies. *Geophysics*, 39(4):441–463.

Telford, W. M., Geldart, L. P., Sheriff, R. E., and Keys, D. A. (1976). *Applied Geophysics*. Cambridge University Press, Cambridge.

Trappe, H., Gierse, G., and Pruessmann, J. (2001). Case studies show potential of Common Reflection Surface stack - structural resolution in the time domain beyond the conventional NMO/DMO stack. *First Break*, 19(11):625–633.

Ursin, B. (1982). Quadratic wavefront and traveltime approximations in inhomogeneous layered media with curved interfaces. *Geophysics*, 47(7):1012–1021.

Vieth, K.-U. (2001). *Kinematic wavefield attributes in seismic imaging*. PhD thesis, Karlsruhe University.

von Steht, M. (2004). The Common-Reflection-Surface Stack under Consideration of the Acquisition Surface Topography — Combined Approach and Data Examples. Master's thesis, Karlsruhe University.

Widess, M. B. (1946). Effect of surface topography on seismic mapping. *Geophysics*, 11:362–372.

Wiggins, R. A., Larner, K. L., and Wisecup, R. D. (1976). Residual static analysis as a general linear inverse problem. *Geophysics*, 41(5):922–938.

Wilson, W. G., Laidlaw, W. G., and Vasudevan, K. (1994). Residual static estimation using the genetic algorithm. *Geophysics*, 59:766–774.

Yilmaz, Ö. (1987). *SEISMIC DATA PROCESSING*. Society of Exploration Geophysicists, Tulsa.

Yilmaz, Ö. (2001a). *SEISMIC DATA ANALYSIS*, volume 1. Society of Exploration Geophysicists, Tulsa.

Yilmaz, Ö. (2001b). *SEISMIC DATA ANALYSIS*, volume 2. Society of Exploration Geophysicists, Tulsa.

Zhang, Y. (2003). *Common-Reflection-Surface Stack and the Handling of Top Surface Topography*. Logos Verlag, Berlin.

Zhang, Y., Bergler, S., and Hubral, P. (2001). Common-Reflection-Surface (CRS) stack for common offset. *Geophysical Prospecting*, 49(6):709–718.

Danksagung / Acknowledgement

Prof. Dr. Peter Hubral danke ich für die Betreuung und sein Interesse an meiner Arbeit und die Übernahme des Referats. Er hatte stets ein offenes Ohr für all meine Fragen. Auch seine Ideen haben die Entstehung dieser Arbeit und die Verbreitung des Programms in der Industrie positiv beeinflusst.

Prof. Dr. Friedemann Wenzel hat das Korreferat für meine Arbeit übernommen, dafür bedanke ich mich herzlich.

Dr. Jürgen Mann danke ich für seine Hilfe bei allen programmtechnischen Fragen. Mein Dank gilt ihm aber auch für die Korrektur meines nicht immer ganz eindeutigen Englisch.

Dr. Thomas Hertweck und **Christoph Jäger** danke ich für das Korrekturlesen meiner Arbeit. Sie haben immer gute Vorschläge gehabt, auch wenn sie selbst unter Stress standen. **Christoph Jäger** danke ich zusätzlich noch für die Unterstützung gerade in der Schlussphase meiner Doktorandenzeit.

Mit **Zeno Heilmann** und **Markus von Steht** habe ich gerne zusammengearbeitet, um die neue Methode auch für die Behandlung komplexer Topographien einzubinden. Auch waren sie eine große Hilfe bei der Bearbeitung der vorgestellten Realdaten.

Erik Ewig danke ich für seine Aufopferung für die Reststatik während seiner Diplomandenzeit. Es war vielleicht nicht immer leicht mit mir, aber zum Schluss ergab sich doch eine konstrukive und freundschaftliche Zusammenarbeit. Somit konnten wir den Grundstein für die CRS-basierte Reststatikkorrekturmethode gemeinsam legen.

Bei **Dr. Franz Kirchheimer** bedanke ich mich vor allem für seine Vorschläge, die maßgeblich an der Entstehung dieser Arbeit beteiligt waren. Auch durch seine Erfahrungen im Bereich der Reststatik konnte ich immens profitieren.

I thank all **sponsors of the Wave Inversion Technology (WIT) consortium** for their support of my work in the last three years. Without their help, the application of the CRS-based residual static correction method to real datasets would not have been that easy.

Furthermore, I thank especially SAUDI ARAMCO for kindly providing the real dataset and giving the permission to publish the results as they have been presented in Section 6.2.

Zu guter Letzt danke ich auch **meinen Eltern**, die mich stets tatkräftig unterstützt haben, so dass ich ohne große Probleme meiner Arbeit nachgehen konnte. Auch **meinen Geschwistern mit Anhang** danke ich für die gemeinsam verbrachte Zeit, die eine Bereicherung meines Lebens ist.

Da ich hier leider nicht alle Personen namentlich erwähnen kann, bitte ich euch, mir nicht böse zu sein und danke allen meinen Kommilitonen und Freunden, die es mit mir ausgehalten haben während der letzten drei Jahre. Die abendlichen und auch nächtlichen Eskapaden haben mir Spass gemacht und die alltägliche Arbeit versüßt.

Lebenslauf

Persönliche Daten

Name:	Ingo Koglin
Geburtsdatum:	17. Dezember 1974
Nationalität:	deutsch
Geburtsort:	Karlsruhe

Schulausbildung

1980 - 1984	Grundschule Leopoldshafen
1984 - 1994	Gymnasium Neureut
17.05.1994	Allgemeine Hochschulreife

Hochschulausbildung

1995 - 2002	Studium der Geophysik an der Universität Karlsruhe (TH)
14.02.2002	Diplom
seit März 2002	Doktorand an der Fakultät für Physik der Universität Karlsruhe (TH)